To Nancy

It was a joy to meet you!

Why? you ask!, may answer — I feel lately sameness of the same routine same predictable routine of daily life — day to day rise and fall of the ~~~~ & sublime to the ridiculous — need experiences to refresh and recharge my the Raison d'être & also change for not!

Samar Majumder
9-9-2014

The Drama of
Man and Nature

Sanat K. Majumder

To my wife, Flora, for her loving support, and to the cherished memory of my parents.

Second Printing ©2009 by Sanat K. Majumder, Northampton, Massachusetts

First Printing ©1971 by Charles E. Merrill Publishing Co, Columbus, Ohio.

No part of this book may be reproduced in any form, electronic or mechanical, including photocopy, recording, or any information storage and retrieval system, without permission in writing for the copyright holder.

International Standard Book Number:
0–675–09224–8

Library of Congress Catalog Card Number:
70–146319

FOREWORD

The intelligent person of today urgently needs a reasoned approach to the problem of man's relation to his environment on earth — an approach in the tradition of the progress of "Western man." The step-by-step recounting of essential requirements of life in its environment, the interdependencies of living organisms, the limits beyond which we may not expect life to prosper — are all narrated here for the reader and in themselves provide *The Drama of Man and Nature.*

Looking at the many new books on conservation and on the environment, one notes a prevalence of emotionalism. When done well, as in *Silent Spring,* with ample documentation, the effect is altogether good. But, unless the goals are clearly defined and the evidence well stated, an author would do well to avoid inciting strong feelings. The literature of man and the environment is also well supplied with proceedings of conferences on conservation matters which often contain gems of wisdom, but mostly are badly integrated and never seem to convey a total picture.

Man needs to face and overcome challenges; his very nature requires puzzles and problems to satisfy his curiosity and stir his imagination. It will be evident to readers of this carefully thought-out and ably documented book that man faces a desperate challenge right now, and a bright future to be bought only at the price of clear thinking and concerted action.

Dr. Majumder rightly places man in a central position in his analysis — not a matter of man triumphing over nature or man occupying a servile role, but finding his proper niche with due regard to his own potentialities and limitations as well as to those of his environment.

One of this book's attributes is its educational value. Its author is a notably fine teacher and we may expect that his book will be used in Conservation and Biology courses as well as read for its own absorbing interest.

Albion R. Hodgdon
Professor of Botany
Univ. of New Hampshire, and
Editor of Rhodora

PREFACE

Environmental biology, as I perceive it, does not deal exclusively with the physical environment, nor is it strictly biological. It permeates nearly every academic discipline. Consequently, it lends itself admirably to the task of exploring a continuity of knowledge in the context of a liberal arts education. This book is written in an attempt to develop such a perspective.

Education is like electricity which turns the blades of a fan and generates wind, passes through a heating coil and emits heat, or burns the filament of a bulb and produces light. The various manifestations of an educated man are attributes of the same driving force. A critical overview of man's "place in the sun" is not only timely but also potentially capable of generating such a driving force among students.

As we view our future needs, the widening dichotomy between the study of humanities and the pursuit of scientific knowledge seems disastrous. Education is a human enterprise and *man* is in the center of it, always recognizing new urgencies and developing fresh sophistication. In the ultimate analysis, it appears, education must include two broad, mutually inclusive areas: (a) the nature of man and (b) nature and man. The ambivalence of man's animality and his capacity for conscious reasoning can be reconciled in such a pedagogic scheme. Our collective knowledge in such disciplines as history, language, social science and philosophy reveals man's nature, his accomplishments and failures, his search for truth. Studies of the physical and natural sciences, on the other hand, help to interpret how man relates to the natural laws and fits into the scheme of nature.

From a few miles up in the atmosphere to several feet under the earth's crust and to great depths of the ocean, lies what we call the *biosphere* where inanimates and animates are huddled in a colossal assembly line of production, consumption and decomposition. An appreciation of these processes as they occur in nature and, also, as they are affected by man's intervention, must be considered as a basic component of education today.

A discussion of man and nature invariably gravitates toward man himself. The rural and urban life of yesterday are fusing to become today's homogeneous industrial society. Man's ecological niche in the biosphere is changing radically. He has released the power of giant rockets and the energy of invisible atoms. He has overpopulated the earth with his own kind. He has conserved, produced and destroyed with great abandon. Various motivating forces are pulling him in different directions—a situation indicative of severely strained perspectives.

Without sacrificing scientific accuracy, the major emphasis in this book is the creation of a healthy overview with which to appreciate the unfolding drama of man and nature. In an effort to accomplish this objective, I have deliberately chosen to straddle the intangible line of demarcation between an esthetic response to nature and the scientific analysis of its various components.

This book is addressed to amateur naturalists as well as to students with ecological orientation. More specifically, it has been designed for non-science majors who wish to develop sophisticated generalizations that are compatible with their assessment of other areas of concentration; the supplementary reading lists at the end of each section have been prepared with this in mind.

I am thankful to many sensitive authors whose thoughts have stimulated mine, to my immediate friends whose empathy provided me with confidence, and to the generation of my son, Tiku, and daughter, Shari, for its constant reminder of urgency. My gratitude goes to Professor Hodgdon for writing the Foreword of this book. Special thanks are due to my wife, Flora, for her endless assistance in the preparation of the manuscript.

S. K. Majumder

CONTENTS

Part I	Retrospect	1
	Living and Non-living	3
	Laws of Thermodynamics, Entropy and Life	6
	"Elixir" of Life: Photosynthesis	7
	Open Systems: Cycles, Feedback and Control	9
	References	11
	Supplementary Readings	12
Part II	Nature	13
	Abiotic Components: Materials	15
	Abiotic Components: Energy	24
	Biotic Components	30
	The Drama of Interaction	42
	References	82
	Supplementary Readings	83
Part III	Man	85
	Mammals and Man	87
	Ancestry of Modern Man	91
	Man's Evolution: Cultural and Biological	94
	Man's Interaction with Nature	95
	The By-Products of Affluence	105
	References	115
	Supplementary Readings	117
Part IV	Prospect	119
	Alone in the Crowd?	121
	Reconciliation	123
	Feast of Knowledge and Famine of Wisdom	125
	Rationalization of Hindsights	127
	Affirmation: A Program of Action	129
	References	132
	Supplementary Readings	133

Part I

RETROSPECT

"The sea lies all about us . . . In its mysterious past it encompasses all the dim origins of life and receives in the end, after, it may be, many transmutations, the dead husks of that same life. For all at last returns to the sea — the beginning and the end."

Rachel Carson

RETROSPECT

Where do we begin? We may begin with the *diversities* of life that distinguish a cow from an oak plant, or with the contrast between a microscopic bacterium and a giant redwood tree that flanks the fog-laden California coast. Or, should we begin with the theme of *unity in diversity* that ties together all biological principles into a few sophisticated generalizations? Since this presentation is intended to possess a synthetic overtone, an examination of some pertinent underlying principles should receive priority. This treatment, then, should set the stage for a more meaningful consideration of the diversities in life (Parts II and III).

As our imagination takes us back a few billion years, an array of current scientific evidence may help us take a closer look into that "primal oceanic soup" where *life* was inaugurated. From there "time's arrow" carried life along on an inexorable path to organization and evolution. Today the man on the street, the wild flower of the valley, and the fish of the ocean are still on the same path together.

LIVING AND NON-LIVING

The difference between living and non-living systems is much too obvious to enter our normal conversation. The need for defining *life* is essentially an outgrowth of man's intellectual and scientific pursuit.

We have come to recognize as living any system that is irritable, has endogenous movements, can metabolize food energy and reproduce itself. Ability to self-replicate or reproduce is probably the *sine qua non* of life, since all life had a precursor that "lived" before.

A cell had remained unchallenged as the basic unit of life until the notorious submicroscopic particulate entities called *viruses* entered the biological scene. Viruses consist of nuclear materials surrounded by a protein coat. Outside living cells, viruses behave as non-living inert bodies of macro-molecules. Most of them have no enzymes and none have metabolism. Within living cells, however, viruses are not only active metabolically but profuse in self-replication. The warm hospitality of the host cell is all that is required to make a seemingly innocuous guest come to life! It would appear, then, that the cell *is* the last frontier of life; and the virus, which assumes a peculiar position between the living and non-living, is the *threshold* of life.

The foregoing description of the virus would suggest that, in the ultimate analysis, the line of demarcation between living and non-living is extremely narrow. In fact, the mechanistic view of life maintains that there is nothing in a living system that, when deciphered completely, cannot be explained in the light of existing laws of chemistry and physics (9; 2). The vitalists, on the other hand, assert that such phenomena as consciousness, behavior and evolution in an organism cannot be explained in terms of mechanistic functions (6; 7). Although we shall return to this intriguing controversy, our present discussion will be limited to empirical observations of what constitutes a living system.

The living world is essentially an organic world characterized by the interplay of complex organic molecules. Spatially, it occupies an infinitesimally small portion of this universe, 99.9999999999 per cent of which is non-living and inorganic.

Scientists tell us that our planet, together with other planets and the sun, was formed from a cloud of cosmic dust under the forces of rotation and gravitation. The original atmosphere of the Earth was reducing in nature, with a predominance of hydrogen gas and its compounds. The relative proximity of Earth, Mars, and Venus to the sun, and their resulting higher temperatures, caused the lighter elements such as hydrogen to diffuse away from their atmospheres; whereas the lower temperatures of Jupiter and Saturn allowed them to retain much of hydrogen and its compounds. This timeless shift in the atmosphere of the Earth can be further recognized by the relatively smaller amount of lighter elements (e.g. hydrogen) that is present today compared to the abundance of non-volatile elements such as silicon.

Ninety per cent of the mass of the universe is believed to be composed of hydrogen, under a constant bombardment of cosmic

RETROSPECT

rays and other ionizing radiation. When the Earth's atmosphere escaped from these hostile elements, the stage was set for the initiation of life on Earth.

Living systems are organic, but obviously all organic compounds are not living. As early as 1820, Friedrich Wohler, and later, Justus von Liebig, succeeded in producing a large number of organic substances from simple inorganic elements. But how did this transition from inorganic to organic forms take place in the primal Earth's atmosphere? That was presumably the major crisis that had to be met for life to come into existence.

We cannot grasp the facts of this remote past but we can indeed reach them through indirect scientific evidence and intelligent speculations (8; 4). We are told that methane, ammonia and traces of carbon dioxide, hydrogen and oxygen constituted the milieu, and water was the "cradle" where the first complex organic molecule was formed. Awesome volcanic eruptions, colossal electrical discharges and endless radioactivity provided the necessary energy for all chemical reactions to take place in the turbulent oceanic "soup." This drama was enacted approximately two to three billion years ago.

The biochemical trial and error then began. It is no longer mere speculation to say that out of the chaos of randomness, organized complex organic molecules evolved. These molecules, individually and collectively, had to take the "fitness" test, as it were, to determine their survival and usefulness in the shifting environment. Proteins and nucleic acids which we now know to be integral parts of any living system—even the most primitive one—must have originated during these dynamic interactions (4; 5). Very recently an exciting suggestion has been made with regard to prebiological protein synthesis (3); experimental evidence indicates that without the intervening formation of amino acids, proteins could have evolved readily in the primitive reducing atmosphere.

There is still one more major crisis to be met before we can pave the way for (a) the inauguration of terrestrial life and (b) the evolution of multi-cellular plants and animals. We shall turn to this momentarily; but first, let us examine briefly the implications of physical laws in relation to the origin and evolution of life on earth.

LAWS OF THERMODYNAMICS, ENTROPY AND LIFE

About 40 years ago a great physical chemist, the late G. N. Lewis, wrote, ". . . living creatures are cheats in the game of physics and chemistry . . . They alone seem able to breast the great stream of apparently irreversible processes. Those processes tear down, living things build up."

Encompassed in the "game" of physics and chemistry are the laws of thermodynamics. The first law states that the total energy of this universe remains constant. To put it another way, energy can be neither created nor destroyed; it can only change from one form to another, the sum total remaining the same.

The second law of thermodynamics relates to the question of *entropy*—the degree of randomness or disorder. It proclaims that the entropy of any process in the universe is doomed to increase irreversibly, the ultimate fate being a state of complete randomness and disorder (maximum entropy). Our universe is, then, like a clock in the process of running down.

In this context, Dr. Lewis' statement gains significance. In a universe which is characterized by increasing disorder, a living system appears to turn the tide by consuming the disorderliness of its surroundings and building an incredible paragon of orderliness within itself. As the sun gets colder, living cells salvage solar energy to build and synthesize large energy-rich compounds with which to carry out intricate functions. The discrepancy created by the living system is, however, quite transient in the scale of time. It is, in fact, more apparent than real. The second law of thermodynamics requires that entropy in any part of the universe may decrease only as long as there is a simultaneous and corresponding increase in entropy in some other part. Order may increase in the organism, provided the environment increases still more in disorder. This, essentially, is what does occur in nature.

Living cells and organisms are open systems; and, under normal conditions, they are continually exchanging matter and energy with their surroundings at a steady rate. In this exchange, the entropy within the cell is at its minimum and that of its immediate surroundings is correspondingly higher. In other words, living organisms choose the least evil—they produce entropy at a minimal rate by maintaining a steady state. This arrangement continues until the organism dies and decays to join the ultimate state of maximum entropy.

In view of the foregoing discussion, it can be said that living organisms also obey the second law of thermodynamics. They *do*, however, "cheat" the law enforcement agency, so to speak, by taking an additional lease on relative stability and organization.

Where did it all begin? The tide was turned when certain specialized cells became capable of capturing solar energy for their own sustenance. As the forthcoming discussion would substantiate, these photosynthesizing cells became our only links to the sun's energy; they constituted the supreme point of contact between the hostile physical world and the docile organic living world.

We should now logically turn to photosynthesis and examine its implications. As Dr. D. I. Arnon points out (1, p. 164): "The understanding of the nature of the energy conversion process in photosynthesis is a part of the intellectual emancipation of man by helping him to understand how the living world has developed in harmony with the laws of physics and chemistry."

"ELIXIR" OF LIFE: PHOTOSYNTHESIS

Love of nature can stem from pure aesthetic considerations, which may or may not precede scientific understanding of the intimate interdependence between plants and animals. In both Greco-Roman and Oriental history, abundant evidence exists to indicate that the riches of Nature were appreciated with reverence. Still today in India, devotional songs and dances accompany the "tree planting" ceremony. The scientific basis of the reciprocity between plants and animals was not established, however, until 1774 when Joseph Priestly, a Unitarian minister from Leeds, England, concluded, "Plants . . . reverse the effects of breathing and tend to keep the atmosphere sweet and wholesome" for the animals.

As we know today, the chlorophyll pigment of green plants captures solar energy and transforms it into chemical energy. During this photochemical reaction, water is "split" and oxygen is evolved. Chemical energy is further utilized to drive the reactions which reduce carbon dioxide to high-energy carbohydrate (glucose) and, eventually, to proteins and fats.

A detailed description of the entire process of photosynthesis is not within the scope of this presentation. We must, however, examine the implications of the process in retrospect as well as in the current context.

It is obvious that, among other macromolecules, the evolution of chlorophyll molecules in the primitive oceanic milieu must have preceded the establishment of an efficient energy-capturing system. Once the solar energy was captured and transformed into chemical energy (food), it could either be broken down (oxidized) to release the energy necessary for metabolic work, or could be stored for sustenance. In other words, living systems had ceased to "consume their own heritage to extinction" and no longer depended on anaerobic fermentation energy alone. The atmosphere gradually became more and more oxidizing due to relentless photosynthesis and the resulting increase in atmospheric oxygen. Oxidative breakdown of food (aerobic respiration) was now possible for the primitive plants and animals. Carbon dioxide, released as a by-product of aerobic respiration, was recycled through the vehicle of photosynthesis.

At the present time, photosynthesis is estimated to convert 150 billion tons of carbon dioxide to sugar and release about 120 billion tons of oxygen per year. To look at the same staggering statistics in another way, photosynthesis involves all the carbon dioxide in the atmosphere once every 250 years and all the oxygen once every 3,000 years!

Photosynthesis, like a storage battery, stores food energy for nearly all living organisms and is a crucial factor in the earth's annual energy budget. In addition, all fossil fuels such as coal, natural gas and petroleum can be traced to photosynthetic activities through early geological periods. It would appear, therefore, that the phenomenal accomplishment of chlorophyll—the "*elixir*" of life—has insured both the continuity and sustenance of life—both the daily bread we eat and the air we breathe.

The atmospheric shift that took place with the advent of photosynthesis had a profound effect on the evolution of higher forms of life. Prior to this biochemical episode, the preponderance of germicidal ultraviolet light, together with the existing reduced atmosphere, limited all life to an aquatic environment. With the relentless release of oxygen into the atmosphere, a layer of ozone formed in the upper atmosphere and became a protective screen against lethal ultraviolet rays. A biosphere which was safer and more conducive to life began to evolve. Through the eons of time that followed, many forms of plants and animals migrated to terrestrial habitats. Higher plants and the phytoplankton of the ocean photosynthesized high-energy carbon compounds and released more and more oxygen. Higher animals developed a complex vascularization and blood system, assuring, as it were, the consumption of excess oxygen.

OPEN SYSTEMS: CYCLES, FEEDBACK AND CONTROL

We have examined how life might have originated on our planet. We have also established a link between living systems and solar energy. However, this relationship of life to time and space fails to answer one essential question: Helplessly placed at the mercy of the second law of thermodynamics which dictates increasing entropy or disorder, how could living organisms develop toward such stability, order and organization? The obvious answer lies in the complex phenomena of adaptation and evolution. Small but significant and heritable changes have equipped organisms, through the ages, to cope with the shifting environment and its extremities. Those who failed to pass this "fitness" test disappeared, a spectacular illustration being the extinction of giant reptiles that lived between 100 and 200 million years ago. On land, and in water and air a similar drama of adaptation was enacted innumerable times and is still in progress (see Part II).

A student of evolution learns that *death* is an insurance for the continuity of *life*. A philosopher or theologian will probably view death as an event *of* life or *in* life. Our present context, however, is limited to the appreciation of the highly evolved process that delays death, the inescapable sanction of nature. This process can be described conveniently under the broad topic of *homeostasis* which refers to the phenomenon of internal self-regulation by an organism, regardless of the external environment.

In contrast with a mixture of reactants placed in a test tube, a living system seldom reaches a true equilibrium. Instead, a steady cycling of material and energy through the organism insures the stability and order of its internal environment. What is more critical, this cycling should, at all times, guarantee orderly gain over the corresponding loss due to consumption. *Senescence*, which is followed by death, occurs when the input and output of material and energy reach an equilibrium.

Cycling of material and energy is not limited to the ensemble of organism and environment. In the course of evolution, this *steady state* has been established between like and unlike individuals and at various levels of organization—population, organismal and cellular.

Theoretically, there can be no beginning to this cycle. There are, however, an inordinate number of factors that can slow down the cycle or speed it up. The mechanism associated with this control is referred to as *feedback* which can be *positive* or *negative*. A thermostatically-controlled room illustrates the phenomenon of feedback; if the room is heated beyond a certain temperature, negative feedback from the thermometer cuts off the heat. To use another illustration, suppose a chain reaction proceeds from $A \rightarrow B \rightarrow C \rightarrow D$ where A is the initial reactant and D is the final product. If D is not utilized or removed rapidly from the reaction site, it will accumulate, and thus provide a message of negative feedback to reactions that precede $C \rightarrow D$. The result may be a retardation in the rate of the entire chain reaction or, in extreme cases, its complete cessation. The picture will become more complex if, in our hypothetical system, D is *directly* connected with A. In any event, it is obvious that a negative feedback system normally tends to stability (e.g. constant temperature in a room). Positive feedback, however, reinforces and even intensifies the applied change which leads to instability of the system.

Looking down on our biosphere from outer space, as many astronauts have done in recent years, we can perceive the operation of a giant cycle that encompasses the non-living inorganic world and the living organic world. By virtue of their photosynthetic capabilities, green plants have assumed the role of *primary producers* in nature. The *herbivores* (deer, rabbit, cattle, etc.) derive their material energy (food) from the primary producers and are logically called the *primary consumers*. The herbivores are, in turn, consumed by the *secondary consumers*, the *carnivores* (fox, hawk, tiger, etc.). Man, the *omnivore*, however, gets the "lion's share" and is capable of utilizing as food both producers and consumers of many types and descriptions. As organisms die, *decomposers*—fungi and decay bacteria—join this carousel of energy flow in nature. The organic remains are thus converted to inorganic fractions that return to the physical world and complete the cycle.

The gigantic cycle elaborated above requires for its operation a steady feedback from many interlocking cycles that exist at various levels of organization. We must appreciate that mechanisms which control the speed and direction of one such cycle might substantially affect many others. For example, the genetic endowment that controls cellular activity is known to be in balance with the micro-environment of the cell and, in turn, with the macro-climate of the habitat. Any disturbance of this balance may ultimately reflect on the population level and, eventually, on the homeostasis of our entire biosphere.

REFERENCES

1. Arnon, Daniel I., "Photosynthesis as an Energy Conversion Process," in Gairdner B. Moment (ed.), *Frontiers of Modern Biology*. Boston: Houghton Mifflin Company, 1962.

2. Crick, Francis, *Of Molecules and Men*. Seattle: University of Washington Press, 1966.

3. Matthews, C. N. and R. E. Moser, "Prebiological Protein Synthesis," *Proc. Nat'l. Acad. Sci.*, 56 (1966).

4. Miller, S. L., "A Production of Amino Acids Under Possible Primitive Earth Conditions," *Science 117* (1953).

5. Oro, J. and A. P. Kimball, "Synthesis of Purines under Possible Primitive Earth Conditions," *Arch. Biochem. Biophys. 94* (1961).

6. Polanyi, Michael, "Life's Irreducible Structure," *Science 160* (June 21, 1968).

7. Schroedinger, Erwin, *What Is Life*. Cambridge, Eng.: Cambridge University Press, 1944.

8. Urey, H. C., "On the Early Chemical History of the Earth and the Origin of Life," *Proc. Nat'l. Acad. Sci. 38* (1952).

9. Wooldridge, Dean E., *The Machinery of Life*. New York: McGraw-Hill Book Company, 1966.

SUPPLEMENTARY READINGS

Blum, H. F., *Time's Arrow and Evolution* (2nd ed.). New York: Harper Torchbooks, 1962.

Bormann, F. H. and G. E. Likens, "Nutrient Cycling," *Science 155* (January 27, 1967), pp. 424–29.

Florkin, Marcel (ed.), *Aspects of the Origin of Life*. Oxford, Eng.: Pergamon Press, Inc., 1960.

Keosian, John, *The Origin of Life* (2nd ed.). New York: Reinhold Publishing Corporation, 1968.

Rosenberg, Jerome L., *Photosynthesis*. New York: Holt, Rinehart & Winston, Inc., 1965.

Wald, G., "The Origin of Life," *Scientific American* (August, 1954).

Part II

NATURE

"I only went out for a walk and finally concluded to stay out until sundown, for going out, I found, was really going in."

John Muir

Wood Duck Pond, Arcadia Wildlife Sanctuary, Easthampton, Mass. (Photograph courtesy of Daily Hampshire Gazette, Northampton, Mass.)

NATURE

Nature has produced an abundance of reasons for life to exist where it does. It is not essential for a nature-lover to know all of these reasons in order to appreciate the riches of nature. In our present context, however, it is not only useful but indispensable that we understand something about the parts played by various components of nature—some under the spotlight of the stage, as it were, and some behind the scenes.

A word of caution may be in order here: a beautiful piece of architecture suddenly falls short of perfection as soon as its building materials—the brick, concrete and cement—are exposed to the viewer and obscure the totality of its intrinsic beauty. Therefore, as we analyze the *abiotic* (non-living) and *biotic* (living) components of nature, the bridges that link them together into an integrated system must not be allowed to disappear from our minds.

ABIOTIC COMPONENTS: MATERIALS

From time immemorial, in both Oriental and Western cultures, earth, fire, air and water have been accepted as the four essential "elements" of nature. Except for the added scientific understanding of each of these "elements," the above generalization remains true today.

As indicated earlier, cyclic transfer of the abiotic components of nature is interlocked with similar operations within living systems. This interlocking insures a steady state of consumption, circulation and regeneration. Three such cycles can be readily recognized: (a) Lithosphere (soil and minerals); (b) Hydrosphere (water); and (c) Atmosphere (gaseous elements).

Lithosphere. The lithospheric cycle of minerals is less apparent since its operation is infinitely slow and evidenced only after the passage of millions of years. Water from the rain or a flowing river carries many dissolved minerals from mountain top to deepest ocean. The rise and fall of land masses resulting from the geologic deformation of the earth's crust transport the same minerals back to different topography, completing the cycle.

Soil is the vehicle for the transport of minerals. As the crossroad of all terrestrial communities, soil constitutes the major substratum for plants and animals on our planet. With the exception of free-floating plants in the oceans, lakes and ponds, all plants depend on soil for their anchorage, water supply and nutrients.

Soil can be conveniently defined as the superficial crust of the earth that has undergone agelong weathering—the combined effects of wind, temperature, water and movement of earth. This weathering results in the flaking, peeling and pulverizing of the parental rock—the inorganic component of the soil. The organic fraction is derived from the decomposition of vegetation by various fungi and bacteria that inhabit the soil.

A barefooted beachcomber can feel under his feet the textural change of soil as he walks from the sandy shore up through various forms of inland topography. Depending on the size of the individual particles, soil ranges from fine gravel (2.00 mm) to silt (0.02 mm) and, finally, to clay (0.002 mm). From a practical point of view, the percentage of silt and clay in a soil sample determines its quality (water-holding capacity, aeration, etc.).

Soil is *residual* when the original parental rock lies directly under the highly decomposed surface layers, or *transported* if it is deposited from elsewhere over a long period of time. The Ozarks and the Appalachians, among other locations in North America, represent residual soil formations. The movement of *transported* soil can be accomplished by gravity (*colluvial*), by running water (*alluvial*), by glacier (*glacial*), or by wind (*eolian*). The deltas of great rivers are formed primarily by alluvial silt and clay. The sand dunes of the Atlantic and Pacific coasts as well as the eastern shores of Lake Michigan are the most common examples of eolian transport of soil. The overlapping areas of the states of Washington, Oregon and Idaho present the most dramatic spectacle of eolian transport. Through eons of time very fine soil particles much smaller than sand have drifted in the wind and accumulated to form what is known as *Loess* which, in places, can be as deep as 70 meters.

As mentioned earlier, the function of soil is to provide water, nutrients and anchorage for organisms that live in it. From this point of view, soil structure is completely correlated with its function. The structure of the soil can be elucidated under four intermingled phases: (a) the solid phase; (b) the liquid phase; (c) the gaseous phase; and finally, (d) the living phase. We have already referred to the solid phase. The essential features of both liquid and gaseous phases are the concern of the agriculturists; the amount of water retained in the soil after the gravitational runoff is probably the most important single consideration in crop production.

The living phase of the soil, which permeates all other phases, plays a vital role in determining soil structure and fertility. Unicellular algae, protozoa and nematodes, as well as earthworms, centipedes and millipedes are the most common organisms found in the soil. It is hard for us to imagine that almost 95 per cent of all insects spend a large portion of their life cycles underground. The burrowing habit of earthworms and the overturning of soil by various other organisms contribute much to the mixing and aeration of the soil.

Hydrosphere. Water is the "cradle of life." One of the basic conditions for life on earth is that water be abundantly available in liquid form. To insure this availability, an efficient cycling of water is in operation which involves interchange of water between ocean, land and air and between the organisms and their environment (Fig. 2-1). With respect to certain land areas, the apparently closed circuit of evaporation, transpiration (loss of water from leaf surfaces), cloud formation and precipitation is often broken by excessive surface runoff and deep seepage. Ordinarily, the demand for water retention in the upper layers of soil is maintained by the thirsty plant roots. Therefore, without the vegetational cover—the watershed—much of the water that reaches the earth by precipitation is lost to the terrestrial organisms. As the untapped water percolates down or runs off to a nearby stream, the parched soil breaks out with unsightly erosion.

The abundance of water in our biosphere is evident both at the environmental and at the cellular level. While three-fourths of the earth's surface is covered by bodies of water, 90 per cent of the weight of a living cell can be attributed to water as well. Living organisms are relatively plentiful as we move away from an area of water deficiency or water abundance and approach an area that the scientists call *interface*—the waterfront, as it were. At the cellular level also, all the vital chemical reactions are known to take

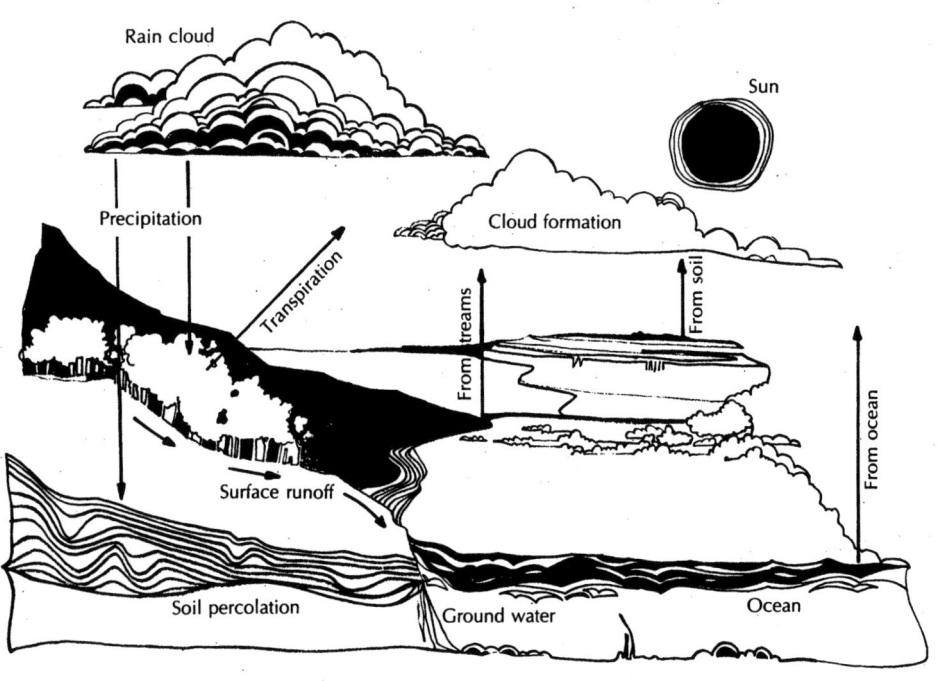

FIGURE 2-1

The water cycle

place in such interfaces where water-soluble and water-insoluble fronts meet.

Water is a universal solvent, permitting a large number of chemical compounds to remain in solution within the living cell. Among other physical and chemical properties that make water relatively unique in the biological world are: (a) great *cohesive force* between molecules and (b) high *specific heat*. The cohesive force that binds molecular layers of water to one another makes possible the ascent of water and dissolved minerals from the roots, by way of the fine conducting tissues, to the aerial parts of the plant. No other single physical force seems adequate to explain how this immense task of water transport is accomplished in such towering trees as the redwoods or eucalyptus.

A considerable amount of heat energy is required to convert water from one phase to another; that is, from solid to liquid and from liquid to vapor. The same statement is true in reverse: large quantities of heat are lost when steam condenses to water or when water solidifies as ice. The ability of water to change readily from one phase to another can be attributed to its high *specific heat*— high capacity to absorb or lose heat with minimal changes in temperature.

The influence of large bodies of water on the adjacent terrestrial climate is of special importance. Ordinarily, the farther north or south we travel from the equator, the lower the average temperature of the region. Yet, the average temperature in January in the Aleutian Islands (latitude 52° N.) is considerably higher than that in Kansas City (latitude 42° N.) during the same time of year. This is obviously due to the tempering effect of the water which cools and heats very slowly and thus affects the diurnal temperature variation in the air mass above the land. The tempering effect is observable also in comparing the coastal region to inland areas of the same latitude. The regions geographically isolated from any large bodies of water are said to have a *continental climate*, characterized by severe extremes of summer and winter temperatures. The *maritime climate* on the waterfront, however, will register a much narrower range of temperatures, both from day to day and from season to season.

Atmosphere. The gaseous exchange between the organisms and their environment is accomplished with the mediation of atmosphere. The major components of the atmosphere are nitrogen (78 per cent), oxygen (21 per cent) and carbon dioxide (0.03 per cent)—a curiously disproportionate mixture of gases. (Water vapor can be considered an additional component, although it is highly variable in density from one locality to another. This variation is obviously due to temperature, relative humidity and the availability of water.) The gaseous interchange in the biosphere can be elucidated by considering three interlocked cycles: (a) nitrogen cycle; (b) carbon dioxide cycle; and (c) oxygen cycle.

Nitrogen Cycle. As noted above, our atmosphere is nitrogen-rich. In view of the fact that most vital cellular components such as nucleic acids and proteins are nitrogenous compounds, it is obvious that living organisms should have great demand for nitrogen. And yet, ironically, no animals and very few plants can use *directly* the enormous atmospheric reservoir of nitrogen. Before it can be utilized in nutrition, molecular nitrogen must be "fixed" in a more negotiable form (nitrate or an ammonium salt). This fixation is accomplished by a variety of microorganisms, including *free-living* soil bacteria and blue-green algae as well as the bacteria which live *symbiotically* in the root nodules of leguminous plants (Pea family). From the foregoing statement, the wisdom of the following agricultural practices should be obvious: (a) addition of decomposed leaf mold to soil; (b) inoculation of rice paddies with blue-green algae—a procedure first adopted and promoted by the Japanese; and (c) rotation of grain crops such as wheat with a legume crop such as alfalfa—a practice referred to as "green manuring" in some Asiatic countries.

Nitrogen removed from the soil as a plant product is consumed by animals as food (Fig. 2-2). A portion of this nitrogen is returned to the soil as animal excreta or dead tissues which, in turn, are subject to microbial action and are once again available to plants. The unused fraction is decomposed to molecular nitrogen.

Carbon Dioxide Cycle. Carbon is the basic constituent of all organic compounds. It enters the organic world when carbon dioxide is reduced to carbohydrate (sugar) through the process of photosynthesis in green plants. The return of carbon dioxide to the atmosphere is concomitant with the process of respiration by living cells. The carbon compounds that are locked in the plant

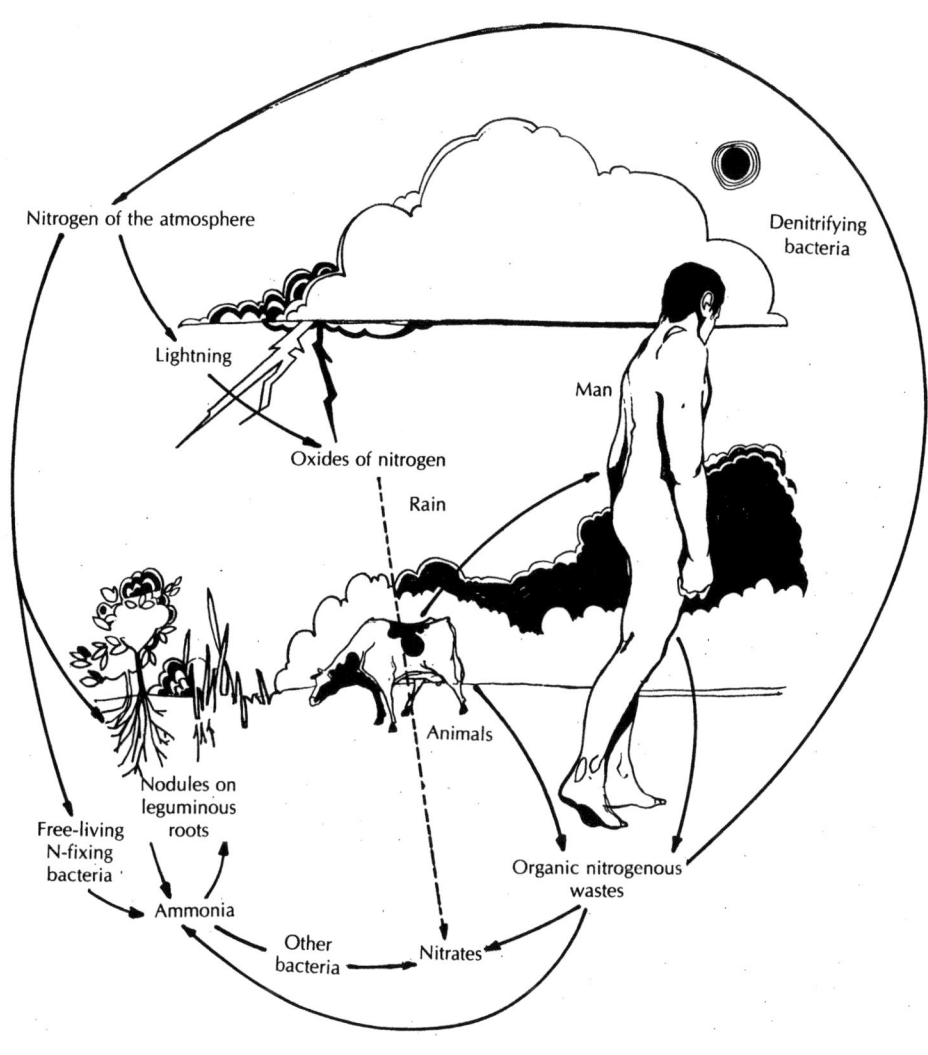

FIGURE 2-2

The nitrogen cycle. Arthur W. Galston, The Life of the Green Plant,
© *1961. By permission of Prentice-Hall, Inc., Englewood Cliffs, N.J.*

and animal wastes become substrates for bacteria and fungi and are eventually broken down to carbon dioxide (Fig. 2–3).

Carbon compounds can take another route when returning to the atmosphere as carbon dioxide. During the early geologic era, a portion of the product of relentless photosynthesis took a detour, as it were, and became incorporated in our earth's crust as fossil fuels—coal, petroleum and limestone (see p. 8). Today, the industrial combustion of these energy-rich fuels is resulting in the completion of this cycle as the smokestacks spew forth carbon dioxide into the atmosphere.

It has already been noted that, compared to nitrogen and oxygen, the percentage of carbon dioxide in the atmosphere is exceedingly low. Because carbon dioxide is directly utilized by green plants, (unlike molecular nitrogen), its replenishment and circulation assume a special significance. Here again, the soil bacteria and fungi play the most critical role by feasting on the dead tissues and releasing carbon dioxide into the atmosphere. As a result, the soil air contains 10 to 100 times as much carbon dioxide as does the atmosphere. This availability, however, does not insure circulation. Ultimately, it is the green plants that "process" large quantities of air for assimilating the small fraction of carbon dioxide. To illustrate this engineering feat, corn plants must find 20,000 lbs. of carbon dioxide in no less than 21,000 tons of air to produce a crop of 100 bushels!

Oxygen Cycle. In the presence of chlorophyll pigment and sunshine, water is "split" and oxygen is released into our biosphere as a by-product of photosynthesis (see p. 7). Free in the atmosphere or dissolved in water, oxygen is utilized by plants, animals and microorganisms in the process of respiration and returned to the atmosphere as carbon dioxide and water.

The most unique feature of the atmosphere, it would seem, is the disproportion and relative stability of its gaseous composition. In a purely physical system this would be considered not only unnatural but essentially impossible. The living world of plants, animals and microorganisms intervenes; all organisms respire and use oxygen; green plants photosynthesize and consume carbon dioxide; some specific organisms take charge of fixing the nitrogen for general consumption. All of these processes, conditioned to different turnover rates and various limiting factors, influence the

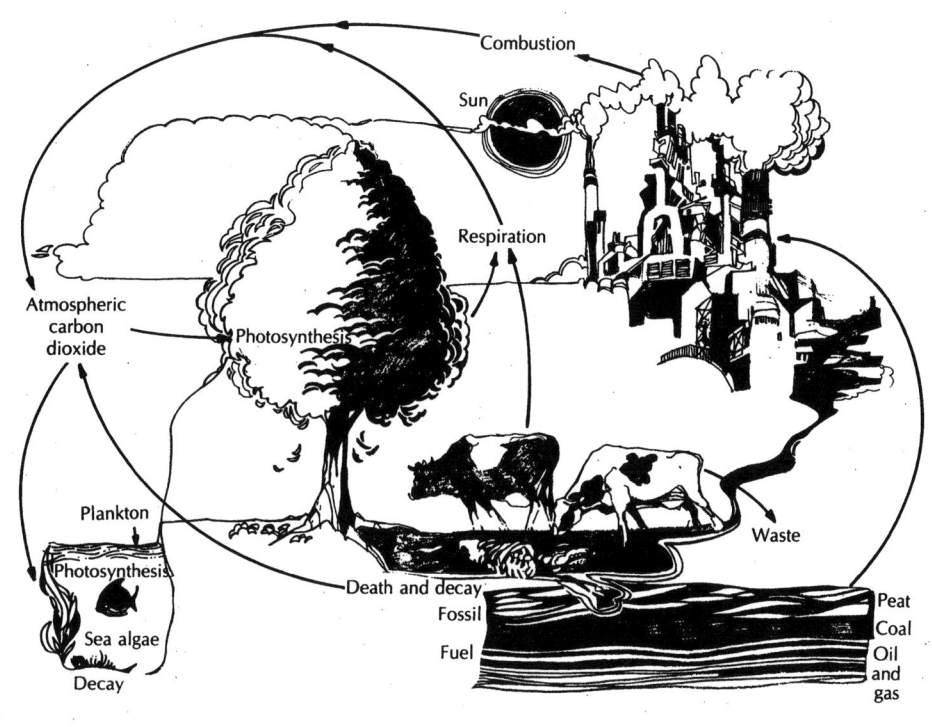

FIGURE 2-3

The carbon dioxide cycle

flow and relative stability of the atmospheric raw materials—the gaseous components. Therefore, the atmosphere today is a result of the physical and the living worlds evolving together as one unit—the biosphere.

ABIOTIC COMPONENTS: ENERGY

The foregoing discussion of the cyclic flow of materials merely recognizes the continuity between the non-living and the living worlds. It does not explain, however, how this continuity is maintained. We shall now examine the driving force—the energy—which must interact with matter to accomplish the orderly procession of intermediate products and their regeneration.[1]

The muscular *energy* of a faithful gardener keeps a garden from succumbing to the emerging weeds. The interaction of heat *energy* with water molecules leads to evaporation and subsequent cloud formation in our hydrosphere. In both instances, energy is utilized to produce a discernible effect (work). In other words, to maintain order in nature—be it in a garden or in the pervasive flow of materials—energy must be "pumped" into the system and work must be done to maintain continuity.

Energy is, then, the capacity to do work. There are two recognizable forms of energy: potential and kinetic. Potential energy is energy in storage, whereas kinetic energy is characterized by action or motion. A toboggan resting on top of a hill possesses a potential energy which becomes kinetic as the same toboggan slides down the hill. In our everyday life, such transformation of energy is a commonplace event: the potential chemical energy of fuel is converted to heat energy by combustion; the mechanical energy of steam turbines comes from heat which, in turn, is derived from the burning of coal.

Chemical energy—food—is the potential energy in any living system. This potential energy becomes kinetic when food is digested and utilized in intracellular activities, resulting in movement, synthesis of new chemical compounds, and growth. To return to

---

[1] According to modern physics, under extraordinary circumstances, energy can be converted into matter, and *vice versa*. For this discourse, however, a distinction between matter and energy will be maintained.

our analogy, the higher the hill, the greater is the kinetic energy released in the toboggan's downward journey. Similarly, there is a hierarchy of cellular macromolecules which store different quantities of potential energy and, therefore, release different amounts of kinetic energy upon digestion. The bigger they are, the harder they fall!

We have just alluded to the basic difference between the transformation of energy in the physical world and that within living organisms. The living system, in order to stay "alive," must store and release energy continuously and not simply convert, explosively, one form of energy to another, as in a fireplace or an automobile engine. To use a redundant but accurate expression, a living system must store energy to release energy, to store energy, *ad infinitum*! The apparently inexhaustible source of this energy is the sun. Our only link with it is, obviously, the green plant which "harvests" the radiant energy and makes food with the mediation of chlorophyll pigment in the presence of carbon dioxide and water. The journey of solar energy continues as the food is passed from mouth to mouth, as it were; from the green plants to the herbivores, and from the herbivores to the carnivores (see p. 10).

We shall now examine more closely the nature of radiant energy — light and heat — that sustains life on earth.

Light. It is not within the scope of this presentation to consider the physics of radiation. We must appreciate, however, that light originates in the bosom of the sun by the fusion of hydrogen atoms and travels endlessly into space as electromagnetic waves. Intercepted and filtered by the outer atmosphere, only a portion of this vast spectrum of radiation reaches our planet. The light spectrum that ordinarily elicits biological reactions extends from the wavelengths of about 400 mμ to 800 mμ including some overlapping areas of ultraviolet and infrared on the opposite extremities of the scale (Fig. 2–4). Almost every object in nature, whether it is the fur of an animal, the petals of a flower, the green leaves of a plant or the surface of a rock, absorbs and/or transmits different parts of the incident light. This selective absorption and transmission of light leave a mosaic of color for the human eye to view and enjoy. The biological implications of this selectivity will be discussed later in this section.

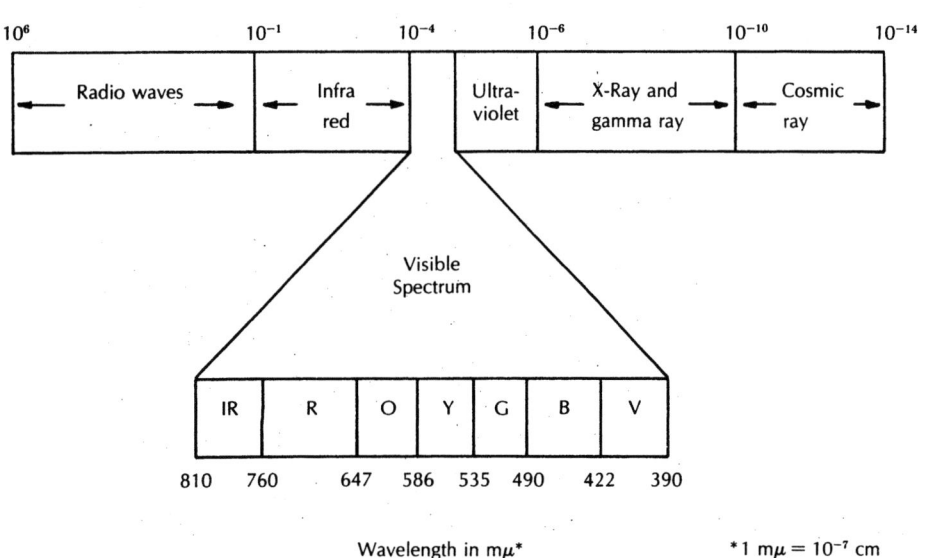

FIGURE 2–4

The electromagnetic spectrum showing wavelength distribution

NATURE

The site of photosynthesis, on land or sea, is the obvious starting point as we consider the energy flow in the biosphere. Chlorophyll pigment absorbs heavily at both blue and red ends of the visible spectrum and transmits much of the yellow and green portions. The photosynthetic rates in relation to various qualities of light also follow the same pattern.

The total energy value of the part of the sun's radiation that reaches the earth is too staggering to be easily comprehended. The "Hiroshima-model" of the atomic bomb was equivalent to approximately 20,000 tons of TNT or 2×10^{10} kilocalories. This represents the total *daily* solar energy incident upon 1.5 square miles of the earth's surface! We must realize, however, that only a little over one per cent of the incident solar energy is used in photosynthesis. The remaining energy is accounted for in a variety of recurring environmental phenomena such as reradiation (loss of heat from the earth's surface at night), and evaporation. How efficient the photosynthetic apparatus must be to do so much for so many with so little captured energy!

Solar radiation is by no means evenly distributed in time or space. There are obvious variations in both quality and quantity of spectral distribution. Biologically, these variations are of supreme importance. Atmospheric interference makes light intensity at sea level about 17 per cent less than at the summit of a mountain with an elevation of 2 miles. With the change in latitude from the equator to the poles, the angle of incident light changes, and its intensity diminishes. The vegetational canopy in the forest causes a distinct vertical stratification of light intensity. As the depth of water increases arithmetically, the intensity of penetrating light decreases geometrically. As a result, under half a mile of ocean water lies more than half the earth's surface, in perpetual darkness!

Equally important are the temporal variations in light intensities, from dawn to sunset, from winter to summer. As will be noted later, the distribution and behavior of many organisms are intrinsically related to this phenomenon.

We have so far considered only light and its interplay in nature. Heat is the other phase of the sun's radiant energy that reaches our planet. The fluctuation of temperature in an object results from absorption or emission of radiant energy by that object. The flow of heat energy, therefore, constitutes a vital abiotic component of nature.

Heat. The intensely hot and hostile crust of the primal earth must have undergone enormous cooling before the life forms of today could appear. Geological evidence in favor of this cooling is substantial.[2] The present temperate regions of the northern hemisphere once supported the growth of tropical plants, as testified by the faithful fossil "fingerprints" of palm trees within one thousand miles of the North Pole. Most geophysicists believe that "polar wandering" of the earth was responsible for these monumental shifts in temperatures: five hundred million years ago the North Pole was near where Hawaii is today; three hundred million years ago it was in the vicinity of present-day Japan. It is also believed that the South Pole migrated from the Atlantic to Antarctica through eons of time.

Regardless of the temperature deviations in our remote geological past, the organisms that inhabit the earth today show a distinct preference for a temperature range in which to function. This range is indeed a narrow one compared to the vast interplanetary temperature gradient (Fig. 2-5). Life, barring any other extraterrestrial form, cannot exist at temperatures below certain minimum and above certain maximum points. Only an efficient cycling of the sun's radiant energy can insure the perpetuation of this apparent *status quo.*

It was noted earlier that the sun's energy is partially filtered by the outer atmosphere before it reaches our planet. The ozone layer absorbs the lethal ultraviolet of shorter wave lengths (less than 290 mμ). At the other end of the spectrum, the infrared is screened by carbon dioxide and water vapor, thereby shielding our planet from excessive heat. Much of the solar energy comes through this screened "window" (5, p. 2) and is redistributed in nature. What better arrangement than this to avoid excessive heat during the day and escape inhospitable cosmic cold at night!

Sandwiched between the ground and the sky, the incoming radiant energy may be absorbed, reflected and/or transmitted by various objects in different proportions. Absorption of visible light by chlorophyll initiates the biological energy flow. On the other hand, the portion of radiant energy which is not biologically fixed follows a strictly physical pathway. Barring any insulation or active

---

[2] See *The Earth.* Life Nature Library. Time, Inc., 1962, pp. 162-64.

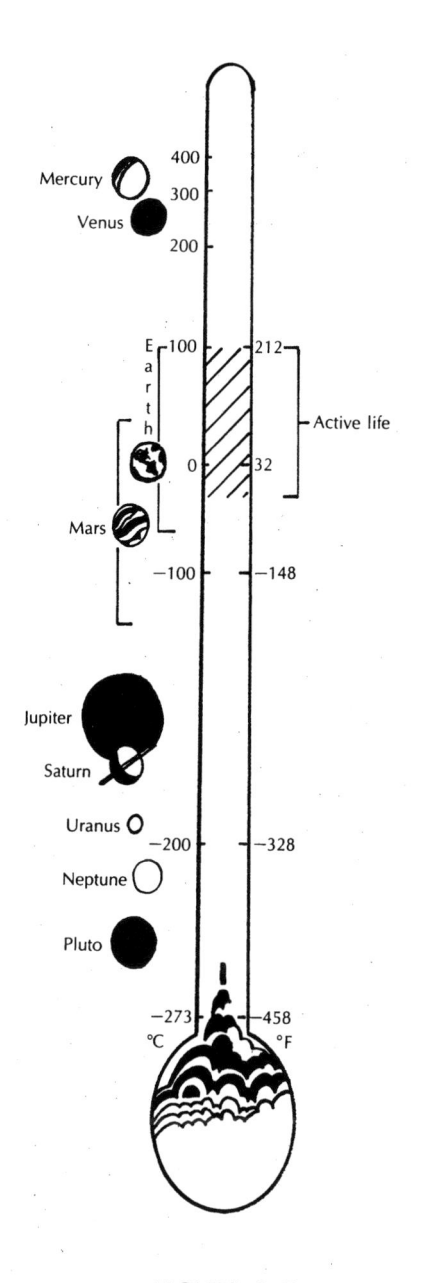

FIGURE 2-5

The planetary thermometer. From Plants in Perspective—A Laboratory Manual of Modern Biology, *by E. H. Newcomb, G. C. Gerloff and W. F. Whittingham. W. H. Freeman and Company © 1964.*

thermoregulation, heat energy must flow from warmer to cooler bodies (infrared thermal radiation). The rate of this flow is obviously controlled by factors such as air temperature, water vapor density and wind velocity. The schematic illustration of heat transfer in nature is shown in Fig. 2-6.

As in the case of light, temporal as well as spatial distribution of temperature also adds to the complexity of nature's energy budget. Within the obviously broad temperature fluctuation between day and night, and between summer and winter, different rates of heating and cooling by soil, air and water invariably produce further gradation of temperature. The most dramatic temperature variations are observed in the intertidal regions where recurring waves bathe the sun-heated beach.

Various factors can affect the spatial fluctuation of temperatures. Among these factors the most important are slope, altitude and latitude. With the sun at an angle of 5°, the solar radiation must traverse more than ten times as great a distance through the atmosphere as it would if the sun were directly overhead. In the mountains, as the elevation increases by 1000 feet, the average temperature decreases about 3° F. Likewise, a gradient of average temperatures is maintained as we approach northern or southern latitudes from the equator.

BIOTIC COMPONENTS

In view of the *modus operandi* of evolution, the environment and the organisms living in it are inseparable. Knowledge of the abiotic components of nature will be meaningless, therefore, unless the biotic components—plants, animals and microorganisms—can be fitted into their respective places under the sun. From the formless mold to complex man, every organism has a habitat in which it is most comfortable. Some have limited niches which they either find by chance or actively seek; others locate territories to which they migrate; still others indulge in the eclectic luxury of building their own homes—their "castles." But all, in order merely to exist, must join the master plan of nature and insure a steady supply of material and energy for themselves.

A thorough, systematic analysis of living organisms is beyond

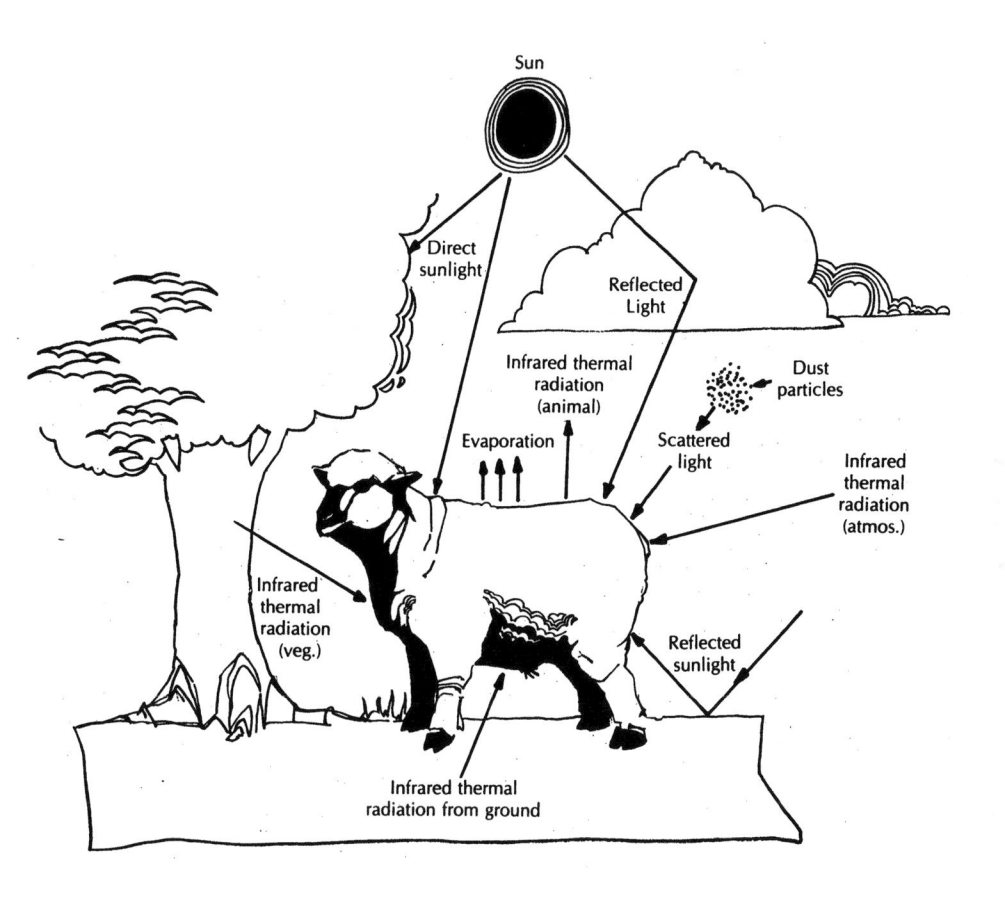

FIGURE 2-6

Thermal radiation in relation to organisms. Modified from The Science Teacher, *Vol. 31, No. 4, 1964.*

the scope of this discourse. An analytical profile of selected groups of organisms will, however, be meaningful. For reasons that will become obvious to the reader, man as a biological species will be discussed in a separate section and not in our general analysis of nature's biotic components.

The process of evolution is probably the most pervasive phenomenon that ties all living organisms together in what has been aptly called an "immense journey" (3). Although controversy exists in determining what constitutes a primitive or an advanced characteristic, it is customary to associate with highly evolved organisms greater degrees of complexity, both in form and function. We shall, therefore, follow this overview in our analysis of microorganisms, plants and animals.

Microorganisms. Plants and animals are easily distinguishable from one another as long as their identifiable characteristics are visible to our naked eyes: green and sedentary redwoods or seaweeds on one hand and the prowling beast of prey or the crawling insects on the other. There are other organisms of innumerable sizes, shapes and descriptions that are visible only under a microscope. Technically, these organisms can be classified under either the plant or the animal kingdom, with some qualifying for both. Their functional characteristics warrant, however, that we identify them in an arbitrary group—the *microorganisms*. For the sake of simplicity, our discussion of microorganisms will be limited to *bacteria* and *unicellular* plants and animals.

Bacteria. A brief reference has already been made to viruses (pp. 3-4). It was noted that viruses constitute what may be called the "threshold of life"; they are non-living macro-molecules when outside living cells but "living" when they take over the metabolic machinery of a host cell.

Unlike viruses, bacteria can be seen through an ordinary microscope and are the smallest of all free-living organisms. Morphologically, bacteria are of three fundamental shapes: rod-shaped, spherical, and helical. Rod-shaped bacteria—probably the most common type—are known as *bacilli*. Spherical bacteria are called *cocci* and helical or coiled ones are *spirilla*.

Ranging from 1 to 10 microns in length and from 0.2 to 1

micron in width, bacteria pack within themselves what may be the barest minimum to stay alive. The bacterial cell does not contain a true, well-defined nucleus. Instead, diffused nuclear materials direct cellular activities. Reproduction is almost exclusively limited to fission or "budding." And yet, special adaptations have enabled these highly diminutive creatures to weather the test of time. By means of their whip-like appendages (flagella), some can swim away in search of moisture. Others produce *endospores*, resistant dense cell-types that can enter dormancy until favorable growth conditions return. In withstanding stress, bacterial endospores are unequalled by the equivalent resistant structures found in many other microorganisms. Even in normal growing conditions, some species huddle together in clusters or chains, ensheathed by a protective capsule.

There is still another remarkable feature that characterizes bacteria: their intense growth rate. A single bacterium is theoretically capable of growing sufficient bacterial cells in one and one-half days to weigh 1000 tons! It is obvious that factors such as competition for food and toxicity limit the growth long before the theoretical maximum is reached. This dramatic potential explains, however, the overabundance of bacteria in air, water and soil, and inside human and animal tissues. Contrary to common belief, only one out of every 20,000 species of bacteria is capable of causing disease in man. Most bacteria, in fact, are not only beneficial to man's commercial interests, but instrumental in the perpetuation of nature's cycles (p. 10). Without the activities of decay bacteria, the death of an organism would be so much more wasteful and final!

Remarkable varieties in nutritional patterns are manifest in the bacterial world, especially with regard to the assimilation of carbon and nitrogen. Most bacteria are *heterotrophic;* that is, they cannot produce their own foods and must depend on external resources. This food may be supplied by the organic and inorganic remains of dead tissues or derived from a mutually-beneficial association with a host (p. 20). Green and purple bacteria, on the other hand, are *autotrophic* and, therefore, make their own food. Equipped with special types of chlorophyll pigments, these bacteria utilize sunlight as their source of energy very much like the higher plants.

Unicellular Plants and Animals. Inseparably merged in nature's enterprise of life and death are numerous unicellular plants and animals. Apparently remote and inconsequential in their habitats, some of these organisms present to the viewers an exquisite array of microscopic patterns and organization.

Fossil records indicate that the unicellular blue-green algae (*Cyanophyta* in Table 2-1) are one of the oldest groups of autotrophic organisms that inhabit our earth today. Like bacteria, these algal cells do not contain any well-defined nucleus. Some blue-green algae, however, have a remarkable advantage over other autotrophs by virtue of their ability to fix atmospheric nitrogen.

A unicellular organism may develop highly complex internal machinery with which to accomplish motility, withstand unfavorable growth conditions and, most importantly, reproduce effectively. We see evidence of such adaptations in the formation of clusters, colonies, or highly intricate patterns of protective outer shells. Among unicellular algae, the cells of diatoms (*Chrysophyta* in Table 2-1) probably display the most ornate designs at a microscopic level (Fig. 2-7).

The intense photosynthesis in the marine environment can be attributed to the numerous diatoms and dinoflagellates (*Pyrrophyta* in Table 2-1) that float or drift near the surface of the ocean. One gallon of sea water can conceivably contain one or two million diatoms. As in the case of other unicellular plants and animals, these algae multiply at an exceedingly high rate by rapid exploitation of the available nutrients. Devoured by animal predators and faced with the problem of dwindling resources, the algal population is soon depleted drastically. Periodically, these natural restrictions are lifted, however; the ocean "blooms" again as the turbulence and wave motions churn up inorganic nutrients from the depths to the warm, illuminated surface.

Unicellular organisms cannot always be assigned to one or another discrete group without qualifications. The Protozoa, for example, are traditionally treated as a part of the animal kingdom. Many protozoan species with appendages (flagella or cilia) have the basic structural features of unicellular algae. The evolutionary tendency in one group, however, is detectably different from that of the other. The unicellular algae have seemingly progressed toward multicellularity without any significant transformation of the individual cells. On the other hand, extraordinary specializations within a single cell appear to mark the achievement of the Protozoa as typified by *Paramecium* (Fig. 2-8).

TABLE 2-1

The Plant Kingdom

Subkingdom: THALLOPHYTA

Name of Phylum	Common Name	Examples
Phylum 1 *Cyanophyta**	Blue-green algae	*Nostoc*
Phylum 2 *Schizomycophyta**	Bacteria	*Streptococcus*
Phylum 3 *Euglenophyta*	Euglena	*Euglena*
Phylum 4 *Pyrrophyta*	Dinoflagellates	*Peridinium*
Phylum 5 *Chrysophyta*	Golden algae	diatoms
Phylum 6 *Phaeophyta*	Brown algae	*Fucus*, kelp
Phylum 7 *Rhodophyta*	Red algae	all true seaweeds
Phylum 8 *Chlorophyta*	Green algae	*Spirogyra*, sea lettuce
Phylum 9 *Myxomycophyta*	Slime molds	*Physarum*
Phylum 10 *Eumycophyta*	True fungi	yeast, mushrooms

Subkingdom: EMBRYOPHYTA

Phylum 1 *Bryophyta*	Mosses, liverworts	Moss, *Sphagnum*
Phylum 2 *Tracheophyta*	Vascular plants	
	Psilopsids (mostly extinct)	*Psilotum*
	Club moss	*Lycopodium*
	Horsetail	*Equisetum*
	Ferns	Christmas fern
	Seed plants	
	Gymnosperm	Pine, spruce
	Angiosperm	Flowering plants
	Dicots	Sunflower, bean
	Monocots	Grass, palm

*Some taxonomists include these two phyla under a separate kingdom, *Monera*, characterized by the presence of *prokaryotic* cells (lacking a well-defined nucleus). Without implying any evolutionary significance and for the sake of simplicity, these two groups have been discussed along with other unicellular organisms and bacteria (p. 32).

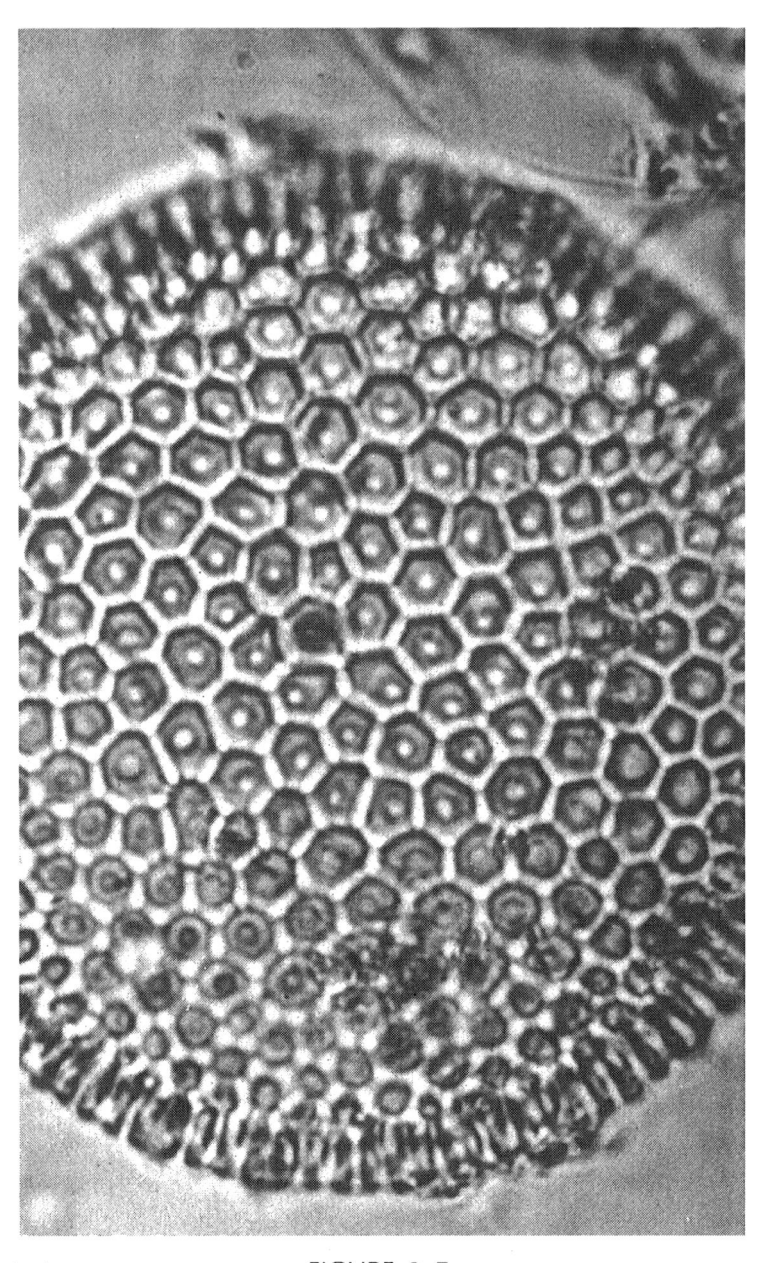

FIGURE 2-7
The ornate structure of a diatom. Courtesy of David A. Haskell.

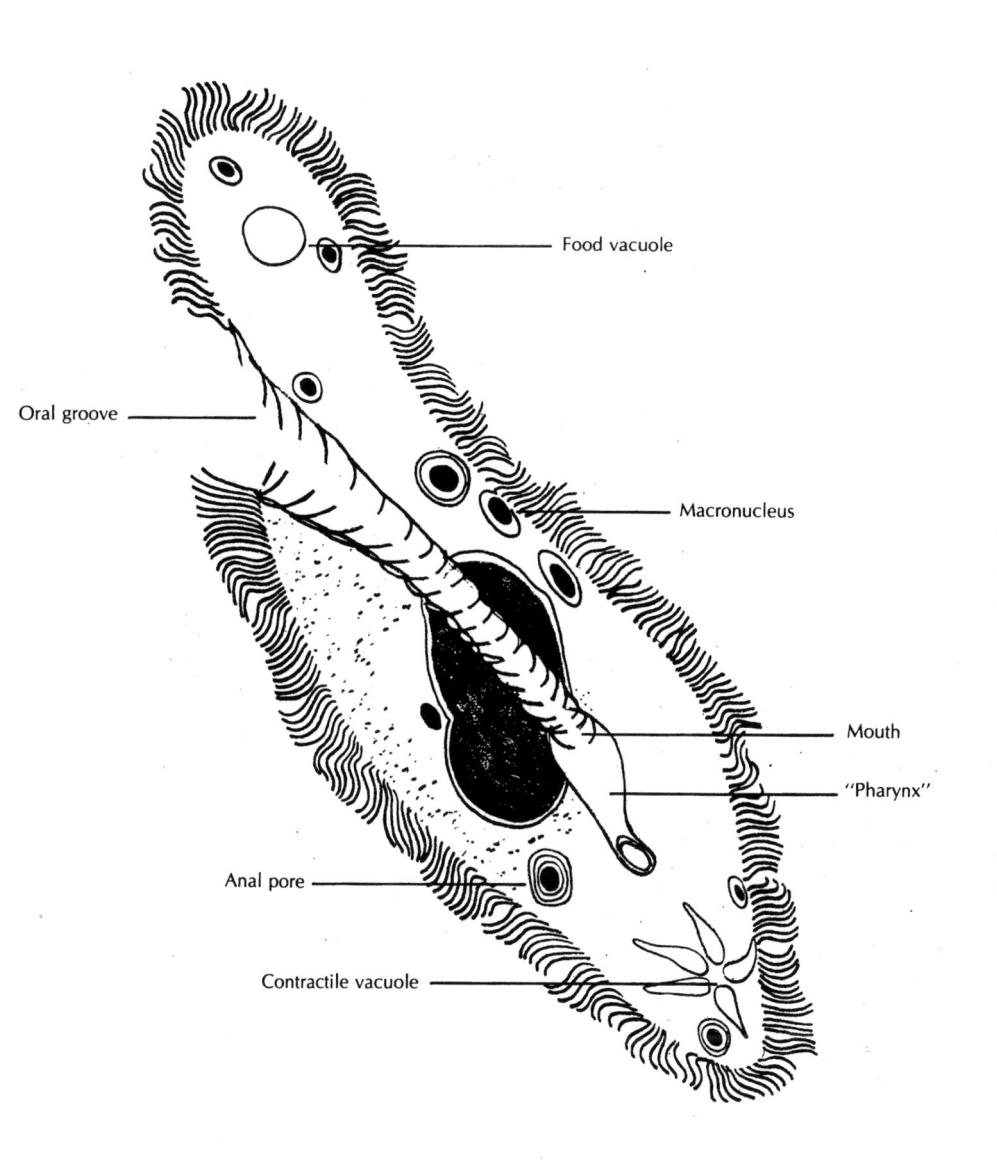

FIGURE 2-8

A diagrammatic view of Paramecium.

Supplied with a few basic nutrients, bacteria can manufacture many other essential foods within their own cells. On the other hand, Protozoa as a group are much more dependent than bacteria on external resources. In nature, these highly heterotrophic organisms, therefore, must find ready-made foods and act as predators on smaller organisms such as bacteria and algae. Thus, the predator-prey relationship, which is so obvious in the natural animal population, actually begins at a microscopic level.

Our discussion of microorganisms would be incomplete without reference to the world of unicellular fungi which includes yeasts. Like bacteria, these organisms are of vital importance to man's commercial interests. They belong to the group *Eumycophyta* (Table 2-1) and are characterized by the absence of mycelia (branching thread-like structures found in "bread mold"). Feasting upon fruits, nectars, and sugar solutions, these unicellular fungi grow profusely by budding or fission and produce ethyl alcohol by fermentation.

Plants and Animals. In view of the foregoing discussion of "microorganisms," it would appear that the evolution of life-forms proceeded from unicellularity to multicellularity. Convincing arguments have been made[3] to indicate that multicellularity is an adaptive strategy; larger size has the advantage of allowing an organism "to be in two places at once," and thereby to be more effective in the utilization of energy and material. In the aquatic medium, organisms are literally immersed in their environment and, therefore, a larger surface area for exploitation of resources appears unwarranted. This probably explains the striking predominance of unicellular and relatively simple forms of life in aquatic surroundings.

From the warm tropical abundance to the chilling arctic desolation, distinct assemblies of organisms grace the terrestrial environment. The higher plants and animals are obviously the results of long evolution. The process of *natural selection* has screened the adaptive strategies of all organisms and brought about innumerable life-forms. In a comparative time-scale of achievement, a

---

[3] William Telfer and Donald Kennedy, *Biology of Organisms* (New York: John Wiley & Sons, Inc., 1965), pp. 92–96.

biologist is often able to label these life-forms as "primitive," "intermediate" or "advanced." In other words, we try to place various classes or subclasses in order of their phylogenetic relationships or lineage.

Plants. Classification of the plant kingdom has undergone periodic changes; a relatively acceptable format is listed in Table 2-1 as the basis for our discussion. Needless to say, no evolutionary sequence is implied in this scheme except to recognize the range of complexities in various forms of plants.

The subkingdom *Thallophyta* includes all plants that are not differentiated into true roots, stems and leaves, and that do not have any vascular tissue to transport water and minerals. With the exception of fungi, slime molds and most bacteria, plants in this group contain chlorophyll as well as other pigments and, therefore, are autotrophic.

As the name suggests, the subkingdom *Embryophyta* is characterized by the formation of an embryo following sexual reproduction. The presence of specialized conducting tissue — xylem and phloem — is the identifying feature of *Tracheophyta* which constitutes the most easily observable component of nature. Ferns, evergreens and flowering plants belong to this group. According to a conservative estimate, there are nearly 250,000 species of vascular plants in the world. Future explorations, especially of the tropical flora, promise to augment this figure greatly.

The success of vascular plants in the terrestrial environment can be attributed to their unique adaptations. To an uninitiated lover of nature, the giant redwood or the vast expanse of prairie grass may not seem to be a display of massive multicellular structures. He can, however, appreciate the "division of labor" among the various parts of the same organism. Multicellularity, as such, is of no adaptive value to these plants except for the specialization of certain groups of cells to function as roots, stems and leaves. Animals move from one place to another either to seek prey or to evade predators. The disadvantage of the sedentary plants is matched only by the advantage of their simultaneous access to light, water and minerals through the deep roots, elongated stems and outstretched leafy branches.

The present overabundance of *angiosperms* — the flowering plants which number 200,000 species — contrasts sharply with the insignificant number (about 500 species) of *gymnosperms*. In the geologic time scale, the last 140 million years were characterized by progressive expansion of flowering plants as the gymnosperms declined. The environmental as well as phylogenetic considerations for such shifts in the vegetational pattern are limited to intelligent speculations and, therefore, are of much academic interest.

Animals. As noted above, maximum utilization of light, water and minerals forms the basis for the adaptation that stabilizes a plant in a particular habitat. The stability of animals, on the other hand, depends on their capacity to "gather food" and their facility to move when chased by predators. In coping with these and other problems of adaptation, the animal kingdom has emerged with a spectacular diversity in size, shape and behavior.

From the gliding ameoba and crawling insect to the flying bird and prowling mammal stretches the panorama of animal life. An abbreviated classification of the animal kingdom appears in Table 2-2, which does not include the subgroupings of any phylum except *Chordata*. Since man, the most highly evolved mammal, is the central theme of this entire presentation, the subdivisions of the phylum *Chordata*, to which man belongs, have been listed in full. It may also be noted that the phylum *Protozoa*, probably the most primitive form of animals, has been included in our discussion of *microorganisms*.

There is much agreement among biologists as to the origin of life in an aquatic medium. About six hundred million years ago, the primitive forms of life probably included some blue-green algae mixed with a variety of simple invertebrates in a marine environment, flanked by lifeless land masses. Fossil records indicate

TABLE 2-2

*The Animal Kingdom**

Name of Phylum	Common Examples
PROTOZOA (5)	*Amoeba, Paramecium*
PORIFERA	Sponges
COELENTERATA (3)	Anemones, *Hydra*
CTENOPHORA	Comb-jellies
PLATYHELMINTHES (3)	Flatworm, Tapeworm

TABLE 2-2 (Continued)

Name of Phylum	Common Examples
NEMERTEA	Proboscis worms
ASCHELMINTH	Hookworm
ROTIFERA	"Wheel animals," Rotifers
BRYOZOA	"Moss animals"
BRACHIOPODA	Lamp shells
ANNELIDA (4)	Earthworm, Leech
ONYCOPHORA	*Peripatus*
ARTHROPODA (6)	Insects (Lobster, Spider, Moth)
MOLLUSCA (5)	Clamps, Snails, Octopus
ECHINODERMATA (5)	Starfish, Sea urchins
CHORDATA	
Subphylum *Hemichordata*	Acorn worm
Subphylum *Urochordata*	Tunicates
Subphylum *Cephalochordata*	Lancelet
Subphylum *Vertebrata*	
Class *Agnatha*	Lampreys
Class *Placodermi* (extinct)	Spiny-skinned shark
Class *Chondrichthyes*	Sharks
Class *Osteichthyes*	Bony fish (salmon, perch)
Class *Amphibia*	Frogs, Toads
Class *Reptilia*	Snake, Crocodile
Class *Aves*	All birds
Class *Mammalia*	
Subclass *Prototheria*	Duckbill platypus
Subclass *Metatheria*	Opossum, Kangaroo
Subclass *Eutheria*	
Order *Insectivora*	Moles, Shrews
Order *Chiroptera*	Bats
Order *Carnivora*	Dog, Cat
Order *Rodentia*	Squirrel, Mouse
Order *Cetacea*	Porpoise, Whale
Order *Logomorpha*	Rabbits
Order *Ungulata*	Goat, Horse
Order *Proboscoidea*	Elephants
Order *Sirenia*	Sea cow
Order *Primates*	Chimpanzee, Man

*Several phyla in the animal kingdom are further subdivided into a number of *classes*, as indicated in parenthesis. Only in the phylum *Chordata*, however, have these subgroupings been listed in full.

that the first invasion of land by amphibians and insects took place between 400 and 450 million years ago. The age of amphibians and reptiles followed. The legendary dinosaurs grazed the land between now-extinct cycads and conifers with no sight of a flowering plant anywhere. The progressive dominance of land by mammals, birds and insects began only between 60-70 million years ago. The "true" man, as a biological species, is only two million years old!

Animals are different not only in their choice of food but also in their manner of eating. The lower forms of animals directly absorb or ingest food particles followed by a minimum of digestion; others have increasingly elaborate digestive systems. Likewise, the development of neuromuscular and respiratory machinery (from gills to lungs) has been characterized by various forms and complexities among the land animals.

The differences between the mammals alone are sufficiently sweeping and can be examined closely. The duckbilled platypus lays eggs, while the young of kangaroo are always born prematurely and find the pouch on their mother's abdomen a comfortable place in which to grow before leaping to independence. Premature childbirth among the placental mammals, however, is usually disastrous. There are mammals that fly and mammals that jump with highly developed hind limbs. There are mammals with flippers instead of legs and others that are quadrupeds. There are mammals that are erect and use their forelimbs for grasping. And finally, there is one who reaches beyond his grasp and dominates the biosphere —man!

THE DRAMA OF INTERACTION

By considering the biotic and abiotic components of nature as individual casts, we can probe into the drama of their interaction. Each drama has a beginning and an end. The drama of interaction in nature, however, stretches between the remotest past and the unforeseeable future; both its curtain time and grand finale, if any, are lost to us mortals!

What is this drama about? It is obvious that plants and animals

FIGURE 2-9

Mammals from down under. Photos courtesy of the Australian News and Information Bureau.

live in their respective niches but *not* independently of one another or of the environment. They compete for space, food and other resources. They differ in the efficiency of their utilization of resources. One generation of organisms lives and dies, and the succeeding generation repeats the performance. This repetition is not as monotonous and uneventful as it may sound, however. The surviving fraction of a new generation may be equipped with certain valuable adaptations with which to confront a shifting environment, to fight off a new enemy or to cooperate with a new friend. This, then, is our drama—the drama of adaptation and evolution. It is dynamic both in time and space; it has a past and a future.

A community of organisms is not merely the sum total of the participating members. Therefore, to an intuitive nature-lover as well as an inquiring biologist, there is another recognizable dimension of nature: the ever-changing complex entity that results from the interactions of the living and the non-living. We shall now examine closely some of these interactions—not only for our intellectual benefit but also to sustain a sense of wonder about the ongoing processes of nature.

Fundamentals of Adaptation. The term *adaptation* has been misused so often that it conveys slightly different meanings to different people. For our purposes, adaptation may be defined as any feature of an organism that allows it to exist and reproduce within its habitat. An adaptation may be physiological, structural or behavioral. The increase in the number of red blood cells is an example of physiological adaptation in response to oxygen scarcity at high altitudes. The fleshy succulent cactus illustrates structural adaptation to the arid desert habitat. The male grouse's elaborate display of plumage during courtship obviously constitutes a highly organized behavioral adaptation.

Did the giraffe "grow" a long neck to reach for the dangling food overhead and thereby "avoid" starvation? From a biologist's point of view, the answer is no. In spite of the substantial correlation that exists between the structure and function of a plant or an animal, an organism does not "evolve" a form to "serve" a purpose. This doctrine of purposiveness or *teleology* is not in keeping with the theory of *natural selection* and *evolution*.

How, then, does adaptation relate to natural selection? Adaptive features accumulate in a population of individuals and many—but not all—of these features are handed down from generation to generation. The environment acts as a sieve to screen the usefulness of adaptations. A greater degree of variability (a larger "gene pool") among the individuals of a population will, therefore, enhance the probability that a particular trait may be selectively inherited in an unstable or changing environment.

Staggering new findings in the field of genetics allow us to examine more closely the question of natural selection. Individuals comprise a population. In order that an adaptive feature may be inherited and represented in the succeeding generation, the individual possessing this feature must reproduce and leave sufficient offspring. However, all individuals in a population do not reproduce; some may be sterile and others may be victims of accident, predation or disease. Under natural conditions, therefore, the existing population can never be a "reincarnation" of the entire constellation of traits possessed by the individuals of the preceding generation. This concept can be accurately summarized by the following statement which is seemingly redundant and evasive: those features of a population that survive to produce offspring *do survive*.

The subject of evolutionary changes, then, is a population and not the pedigree of an individual. The adaptations of an individual are consequential insofar as they belong to the reproducing fraction of the population; they are useful insofar as they impart to the surviving offspring a gamut of variability from which future parents emerge.

All heritable adaptive changes are, by definition, genetic in origin. For the reasons discussed above, however, all changes are not inherited; and the changes that *are* inherited are not of identical survival value. An inherited trait may sometimes remain undetected in the population as a *preadaptation*. This trait is neutral at the time but manifests positive or negative survival value if substantial changes occur in the environment. A tropical vegetation, for example, may contain in its population characteristic traits for cold-tolerance which are neutral features from the standpoint of natural selection in a warm climate. If human response to weightlessness is in any way controlled by his genetic endowment (preadaptation), we shall find out only when the reactions of many orbiting astronauts are studied statistically. Who knows what capacity for interplanetary travel we already harbor among us until we allow ourselves to be "selected" by an extraterrestrial environment!

Specific Adaptations. Placed in the same habitat and environmental conditions, organisms inevitably interact with one another. The results of such interaction vary, however. We see the oak and the pine compete for dominance. Cooperation between some fungi and algae is vital in such a "joint enterprise" as lichen formation. Squirrels and acorns benefit from each other's association although neither depends on the other for survival. The branches of live oaks provide unselfish aerial support for the hanging Spanish moss. Just by "throwing their weight around" buffaloes and elephants trample the undergrowth more than they partake of it. And finally, we see organisms that must kill others for food. These recognizable forms of interaction have been listed in Table 2–3. Needless to say, this list is a relatively simplified version of more complex phenomena that occur in nature.

Every organism has a preference for a distinct combination of environmental factors (e.g. water, temperature, light) in which it functions with the greatest efficiency. This condition is called the *optimum*. As a corollary to this statement, it is obvious that the same organism has *minimum* requirements below which—and *maximum*, above which—its function cannot be observed. What may be a minimum factor combination for one organism, however, may be optimum for another.

From the forest through the grassland to the desert, environmental factors provide a spectrum of combinations to which various plants and animals are adapted. The forest floor is relatively dark due to the canopy of towering trees, moist due to water retention by plant roots, and cool due to the vegetational insulation. As we emerge from the forest, the light intensity, temperature and moisture change drastically as does the soil quality. New combinations of factors prevail on our route to the desert which climaxes an extreme condition.

What is indicated above is a horizontal stratification of factor intensities typical of terrestrial habitats. In the mountains as well as the ocean, the stratification is vertical instead. It may also be noted here that it is futile to attempt to isolate individual factors in a complex interacting system. It is, however, possible to recognize

TABLE 2-3

Two-species Population Interaction

Type of Interaction	Nature of Interaction* A	Nature of Interaction* B	General Result of Interaction	Example in Nature
Neutralism	0	0	Neither population affects the other	Rabbit and squirrel in forest
Competition	−	−	The more affected population is eliminated	Oak replaces pine
Mutualism	+	+	Obligatory and beneficial to both	Root nodule bacteria in legumes
Protocooperation	+	+	Beneficial to both but not obligatory	Squirrel and acorn
Commensalism	+	0	Obligatory to A; B is unaffected	Spanish moss on a tree branch
Amensalism	−	0	A is inhibited; B is unaffected	Buffalo trampling on undergrowth
Parasitism	+	−	Obligatory to A; B is affected	Plant-animal diseases and Predator-prey relations

*+ Growth of population increased
− Growth of population decreased
0 Population growth is unaffected

Adapted from Eugene P. Odum, *Fundamentals of Ecology*, 2nd ed. Philadelphia: W. B. Saunders Company, 1959, p. 226.

in certain types of habitat the predominance of one factor over the others. With this general outlook, let us now examine some specific adaptations in relation to (a) water, (b) temperature and (c) light.

Water. Some parts of northeast India and the island of Kauai in Hawaii record as high as 500"–600" of annual precipitation. The rainfall in the arid southwest of the United States, on the other hand, rarely exceeds 5"–15". Various degrees of scarcity or abundance of available water have punctuated the earth's surface with conditions that extend from desert to tropical forests.

Arid regions are as diverse as the climatic and geological factors that produce them. Long rainless summer, drying wind, wasteful and unpredictable torrential rain, hot day and cool (sometimes frosty) night—these are some of the strange conditions that prevail in the desert environment. Yet, the seemingly inimical and desolate expanse of desert sand and rock does support life, which is surprisingly abundant and varied.

Plants that are prevalent in arid regions are generally referred to as *xerophytes*. Taxonomically, these plants are not related. Structurally, they vary from the giant cacti to inconspicuous liverworts and resurrection ferns. These plants have one thing in common, however: they are well adapted to a drastically insufficient water supply. Three types of drought adaptations are recognized among desert plants: (a) drought *tolerance,* (b) drought *avoidance,* and (c) drought *evasion.*

Plants *avoid* drought by storing sufficient water in their succulent body structures (e.g. cactus), by reducing the rate of water loss (e.g. closure of stomata during the day) or by reaching with long roots for the deep underground water source. The drought *evaders,* on the other hand, remain quiescent during the driest part of the year and come to "life" to complete their life cycle hurriedly during the very short rainy season. Many fungi, bacteria, liverworts and mosses belong to this group. Other noteworthy examples of this unique type are some diminutive herbs such as desert dandelions. Their seeds accumulate certain germination inhibitors which must be washed away by a minimum of rain water before they germinate and grow—a kind of "chemical warfare" between the organism and its hostile environment (14, p. 137). In contrast with

FIGURE 2-10

Saguaro cacti formation in the arid regions of Texas. Photograph by John R. Meyer.

the above two groups, drought-*tolerant* plants have the cellular and metabolic machinery to withstand the extreme stress of dehydration without sustaining any permanent injury. The creosote bush (Fig. 2–11) is probably the most drought-resistant higher plant.

The "deathly quietude" that prevails in the desert at high noon is deceptive. The untrained but curious eye may soon meet a multitude of animals that crawl, burrow, dart or fly. Insects are the predominant creatures in a desert habitat. These herbivores fall prey to carnivores and, eventually, join the food chain that may end up in a coyote or a ringed-tail cat. Obviously, organisms such as lizards, snakes, kangaroo rats and birds also intervene to claim their shares in the food chain.

The biological problems confronting desert animals are essentially the same as those experienced by all other terrestrial animals; however, in the desert these are more acute. Efforts to avoid desiccation have produced a myriad of behavioral adaptations among these animals. What could be better protection against water-loss than to present to the inhospitable environment an impervious skin, to live on dry or nearly dry food, to reduce activities intermittently (e.g. *diapause* in insects), or to excrete little or no liquid?

The insects and many other organisms confine their activities to the hours of darkness when the temperature is low. Others burrow into the sand, a behavior which, alone, may be sufficient to guarantee survival in arid climates. Snakes and lizards slither into rock crevices for shade and protection.

Although far less numerous than invertebrates, several vertebrates have become well adapted to arid regions. The number of mammals is largely limited by the scarcity of food during the long dry season and by the fact that their rate of breeding is too slow to permit a rapid response to a sudden abundance of ephemeral vegetation.

In contrast with deserts and relatively arid regions, the deciduous, evergreen or tropical forests testify to abundance of water. The composition of these forest communities varies a great deal, however, as we follow a temperature gradient from the tropics to the temperate regions. The competition among organisms for water becomes comparatively inconsequential once the forest formation is complete; other factors such as nutrition, light and humidity play a greater role in determining the profile of biotic interactions.

The tropical rain forest deserves a special mention due to the

FIGURE 2–11

Creosote bushes growing in the arid regions of Texas. Photograph by John R. Meyer.

overabundance of life there. Mosses, ferns and many luxuriant herbs dominate the forest floor. Short woody shrubs, adapted to low light intensities, occupy the next level under the shade of tall trees. Orchids and various epiphytes find support on the moist trunks and branches. Occasionally, an ungrateful strangler fig is in the process of choking its host with a tight embrace while the lianas or woody vines—strong enough to withstand a "swinging Tarzan"—dangle from towering treetops.

Animal life in a tropical forest is profuse. According to one estimate, more than 80 per cent of bird species are of tropical origin. The awe-inspiring dissonant chorus of birds has now become a stereotype of the brooding jungle. Thriving in the constancy of warmth and humidity are frogs, toads and other amphibians, some spending much of their lives in the trees. Bees, beetles and butterflies exploit the greenery and reproduce prolifically. These herbivores, as do those in the desert, satisfy the appetite of the small carnivores.

Mammals appear to be more scarce than they are. Many are camouflaged in the foliage and others prefer the nighttime for their activities. This nocturnal habit is not designed to avoid high temperatures, as in the desert, but to evade large predators and to seek other nocturnal neighbors as prey. The monkeys seem to be the exception; they choose the daytime to monopolize the food resources, undisturbed by too many competitors.

It has been noted frequently that migratory birds break their long flights near a body of water where there is an abundance of food. The aquatic plants (*hydrophytes*), whether totally submerged or free-floating, provide sustenance for a great many other animals (e.g. muskrats, snails, ducks). These plants themselves are examples of curious adaptations which include *heterophylly* (Fig. 2-13), reduced oxygen requirement, and formation of a long slender stem (e.g. water lily) for anchoring in the silty bottom. Some of these habitats, however, are characteristically rapid in changing both their vegetational composition and animal life; these changes occur as the water becomes progressively shallower due to deposits of plant and animal remains (see p. 66).

Certain varieties of plants are well adapted to the unpredictable increase in water level due to floods. In India and some parts of Southeast Asia where incessant monsoon rain can often cause floods, these plants are known to possess the adaptive feature of

FIGURE 2-12

Tropical rain forest showing a rare species of "tree fern" in the foreground. Photograph by John R. Meyer.

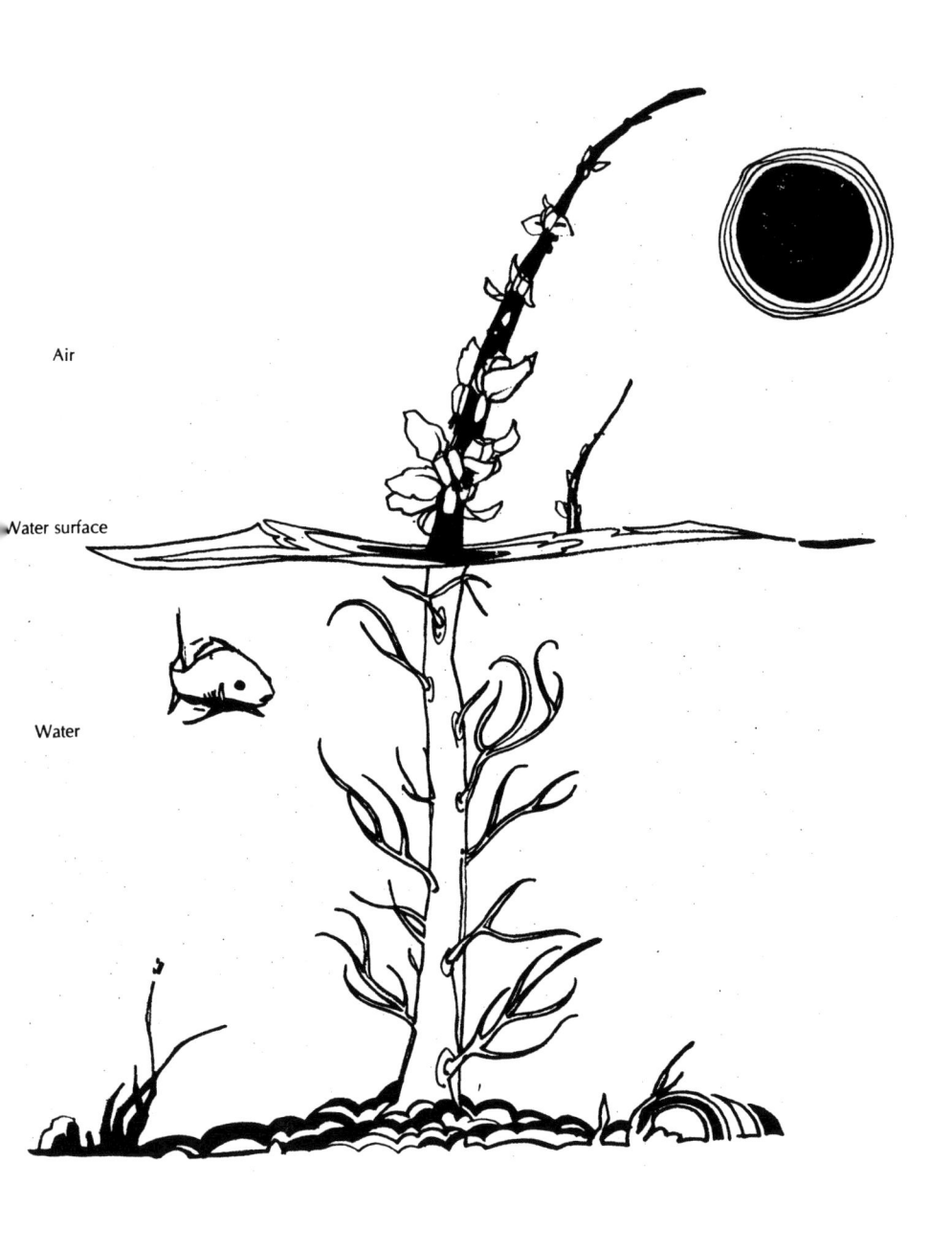

FIGURE 2-13

Heterophylly. A diagrammatic view of a fresh-water plant with two kinds of leaves.

growing "with the flood water." There are some varieties of rice that exhibit this characteristic rather dramatically. Ordinarily, rice plants do not exceed 3–5 feet in height. Under flood conditions, the same plants will grow 10–12 feet in height with their fertile stalks of grain safely above the rising water surface at all times. The farmers harvest them from canoes!

In spite of an abundance of water, plants may suffer from drought if their roots fail to absorb the water. This situation can be construed as "physiological drought" which exists in the inland salt lakes or in the saline coastal regions. Plants that are specifically adapted to survive high salt concentrations are known as *halophytes*.

Considering the fact that the salt concentration in the Great Salt Lake of Utah is as high as 28 per cent, it is impressive that more than two dozen different organisms are known to live there. The presence of succulent halophytes in the coastal areas testifies to the struggle for water. On the tropical seacoast where muddy flats are periodically drenched by the tides, we may find still more specialized adaptations for coping with salinity. These habitats are generally populated by the *mangroves*. Unstable physical support and great limitations of oxygen supply in the root zone are the two major problems that most mangrove plants must face. The *stilt roots* that emerge from the base of many tree trunks provide additional anchorage for the trees. *Pneumatophores* or breathing roots are also common features of many plants in this habitat; reminiscent of the "knee roots" of Florida's swampy cypress, these roots project vertically into the air from the vast network of the submerged horizontal root system.

Seeds of mangrove trees will invariably fail to germinate if they are cast into the soft, highly saline mixture of mud and water underneath. This predicament is solved as plants acquire *viviparous* germination. With this fantastic adaptation, seeds germinate while they are still inside the fruit, aloft in the tree. After the embryo and the seedling have grown sufficiently large, the young plants may be dispersed without the risk of a muddy, asphyxiating burial.

Temperature. Biologists argue that the kind of life that exists on earth could not conceivably evolve on other planets where, among other factors, intense heat and sub-freezing cold prevail. From this cosmic viewpoint, the temperature range of 30°–120° F, in which active life is manifest, is incredibly narrow indeed. The adaptation of various organisms to this limited temperature range is further complicated by the staggering fluctuations that occur from day to night, season to season, sea level to mountain top and finally, from one latitude to another.

The opposite extremes of temperature adaptations testify to a colossal ongoing process. There are algae that find a hot spring (75°–77° C) just the right place in which to grow and multiply. Others occupy certain arctic areas where the temperature rarely goes above the freezing point.

An organism's need to adapt to various temperature ranges is more pronounced in a terrestrial habitat than in the relatively shielded aquatic medium. In open atmosphere, all organisms are subjected to solar radiation which also is the sole source of thermal energy. We noted earlier that part of this energy is stored by the process of photosynthesis, and the remaining portion is dissipated in various ways such as reradiation and reflection (p. 27). The fact of the matter is that if the absorbed solar energy is not utilized biologically, it will generate heat; and if the heat is not removed by some effective means, the body temperature of the organism concerned will rise—at times, disastrously. The reverse of this statement is also true: if the organism fails to retain heat when it is needed, or loses more than its basic metabolism will allow, the consequences can be equally undesirable, if not fatal.

Plants adjust to environmental temperature variations very differently than do animals. Green plants absorb very efficiently the bulk of solar energy from the ultraviolet and visible range of the spectrum. The absorption falls off dramatically, however, in the near infrared (thermal) region where the incident energy is most intense. In other words, "plants absorb efficiently where they require the energy, absorb poorly the near infrared to keep from becoming overheated" (5, p. 2).

In spite of this selective absorption of solar energy, the heat load of green leaves can be considerable at times. The evaporation of water from the leaf surface (*transpiration*) constitutes a natural cooling system which is further aided by wind convection. In addition, the respective activities of sunlit and shaded leaves—their photosynthesis and transpiration rates—help keep the heat transfer in balance. On a sunny summer day, the photosynthesis in overheated leaves can be much impeded and this is compensated for by shaded leaves which register a more favorable temperature. On the other hand, "the pattern of energy transfer on a cool summer day would favor the sunlit leaf and prevent the shaded leaf from reaching the optimum temperature range for photosynthesis" (6, p. 83).

Life is equally vulnerable to extreme cold temperatures. Adaptations to cope with this predicament are dramatically evident in the temperate, arctic and antarctic zones of our planet where daily and seasonal temperature variations are drastic.

Faced with unfavorable environmental conditions, an organism may enter a state of dormancy and resume its active phase as favorable conditions return. We have noted how the desert annuals "hold off" growth and await the welcome rain. With respect to cold temperature, the phenomenon of dormancy provides the single most effective adaptation in the plant kingdom. The environmental cues associated with the fall and winter trigger off within the plants the necessary physiological changes that enable organisms to combat the freezing temperature. What is, however, more critical than dormancy itself, is the timing or synchrony with which this protection mechanism sets in. Obviously, the manifestation of dormancy must coincide with the onset of cold temperature.

The above contention may be best exemplified by the dormancy of seeds, bulbs, and buds. With remarkably precise timing, the seeds of many annuals, the buds of deciduous trees and the perennial bulbs (e.g., daffodils, hyacinth, crocus, tulip) enter dormancy every year during the autumn. This is not just a temporary quiescence, but an "anticipated" state of dormancy which is a heritable adaptive feature conferred on the organism by natural selection and evolution. The "fake" warmth of a "January thaw" will not break the dormancy of wild crocus! To be more accurate, those crocuses that are "fooled" by the January thaw will never be seen in the population of spring flowers.

Although the mechanism of dormancy is a subject of much scientific investigation, its true adaptive significance is rarely appreciated by the untrained observer (1, p. 28). It has been suggested that the embryos or seed-coats of the seeds may remain relatively impervious to water during their dormancy. Some dormant plant structures are known to contain germination-inhibitors which gradually disappear by spring or early summer. In the bulb plants, the flower that emerges in spring is actually initiated, if not fully formed, during the preceding summer, but is carefully "tucked away" in folds requiring the intervening winter to unfold.

With higher altitudes and latitudes, the average temperature of a region drops proportionately. This global temperature gradient has a significant effect on plant distribution (Fig. 2–14). A mountaineer climbing a mountain in the tropics may find on his way to the peak the same type of vegetational changes that he would find if he followed the vegetation latitudinally from the equator to the polar regions.

There is one noteworthy feature about the altitudinal distribution of vegetation: because the south-facing slope receives a greater amount of solar radiation, it is considerably warmer than the north-facing slope. This results in, among other observable differences, the formation of a tree-line which is substantially lower in altitude on the north-facing slope than on the south-facing slope.

Extreme temperatures affect animals in a variety of ways (Table 2–4). Equally varied are the means by which they combat thermal fluctuations in their habitat.

Animals such as mammals and birds are known to be *homeothermic* (warm-blooded) because they possess what one might call physiological thermostats, maintaining a relatively constant body temperature irrespective of the thermal change in the environment. The average body temperature of most mammals is between 96.8° and 102.2° F, whereas that of most birds varies between 104° and 109.4° F. To illustrate this extraordinary physiological adaptation, one only needs to mention the Eskimo dogs that maintain a body temperature of 100.9° F in spite of a prevailing external temperature as low as −22° F.

Insects, fish, amphibians, and reptiles, on the other hand, have no physiological machinery for internal temperature regulation. These animals are *poikilothermic* (cold blooded) and therefore, their body temperature corresponds very closely to the external temperature in which they live. Aquarium lovers appreciate that the temperature of the water must be critically maintained in order for tropical fish to feel comfortable in their artificial habitat.

The success of homeothermic animals in a cold climate can be attributed to (a) their ability to produce heat by metabolic breakdown of food and by enhanced physical activity, and (b) their ability to retain this heat when needed or to lose it when they must. Arctic mammals produce a heavy coat of fur and fleece. They store

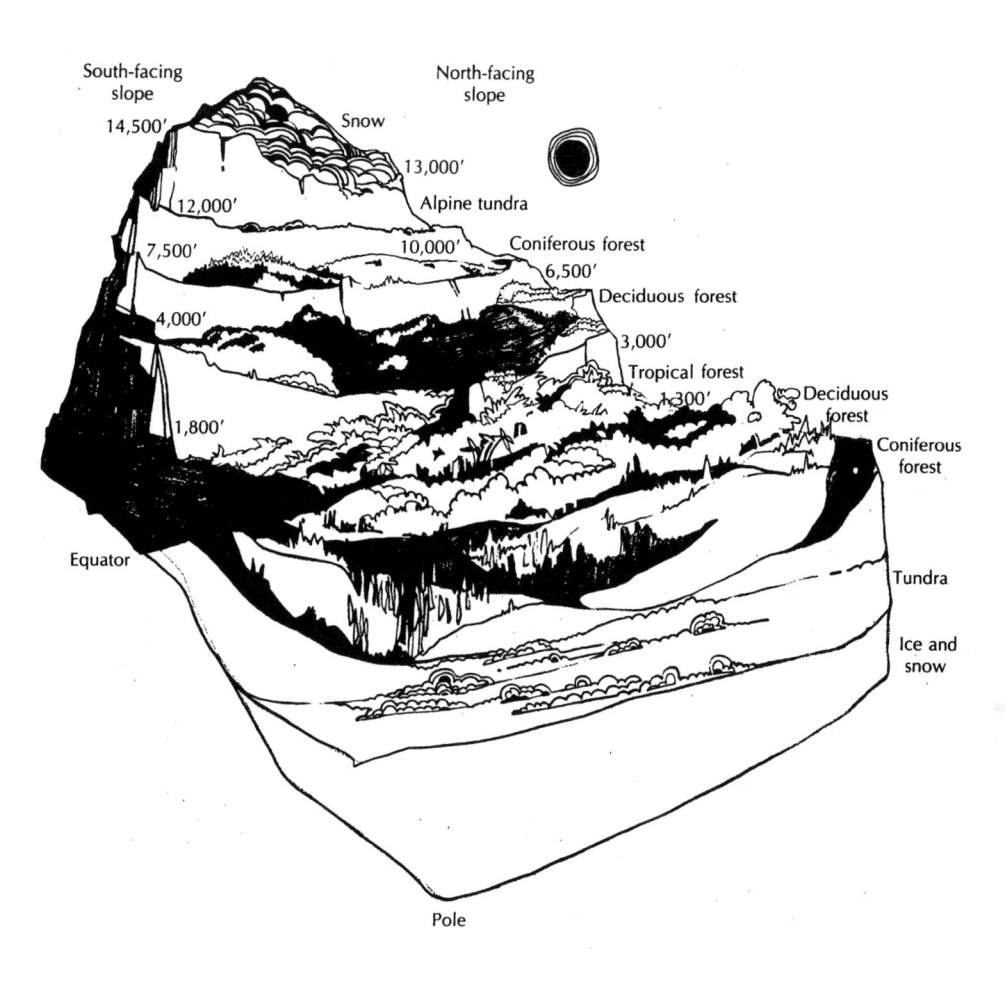

FIGURE 2-14

Parallel between latitudinal and altitudinal distribution of plants. Note that the vegetational distribution on the warmer south-facing slope is different from that on the north-facing slope.

TABLE 2-4

Survival of Vertebrates in Cold Temperature
(from the Handbook of Biological Data, 1956)

Group	Organism	Lower Temperature Limit °C (LD_{50})*
Homeotherms		
	Large arctic mammals	−40
	Rabbit	−35
	Cattle	−13
	Rat	−35
	Duck	−40
	Pigeon	−40
	Sparrow	−30
Poikilotherms		
	Alligator	−2
	Garter Snake	0
	Wood Frog	2.5
	Common Toad	−1
	Coho Salmon	0.2
	Channel Catfish	0

*LD_{50} indicates the temperature at which 50 per cent fatality is expected.

food as fat for slow digestion. Feathers provide insulation for the birds.

What if the external temperature rises to the point that the animal must lose heat to remain comfortable? Loss of heat by animals can be both active and passive. Men and horses sweat, dogs pant, and cats lick their bodies to cool off. Seals must wave their extremities — the flippers — in the air to dissipate heat as the remainder of their fat body is well insulated for swimming in cold arctic waters.

The body weight and the metabolic rate of an animal are inversely related. A small animal such as a rat has a larger surface/volume ratio than an elephant and, therefore, is likely to lose heat to the environment at a faster rate. Clustered together, small animals are known to share "communal heat." This is not sufficient, however. Faced with the problem of heat retention, these disadvantaged animals must consume more food in order to accelerate their metabolic rate. A shrew—one of the smallest living mammals, weighing not more than 4 grams—must eat its body weight of food every day!

In contrast with homeotherms, the poikilothermic animals, by virtue of natural selection, are limited to a narrow habitat where temperature fluctuations are not drastic. These animals are endowed with efficient behavioral patterns to combat temperature extremes (8, p. 162). Turtles bury themselves in the silty bottom of the pond. Reptiles and insects orient themselves to sunshine in order to increase their body heat on a cool day. Butterflies are known to flutter their wings violently before launching into the ominous cold air. To avoid the heat of the shimmering desert sun, snakes seek burrows and insects crawl into shaded niches; the cool quiet of the night brings them forth in search of food.

The most unique form of thermoregulation by an organism is to require none. This situation occurs when a plant goes into dormancy, animals hibernate, or insects undergo *diapause*. Small mammals such as ground squirrels or woodchucks overwinter or surmount other unfavorable seasonal conditions in a state of suspended animation. With shallow breathing and greatly depressed heart beat, they allow their body temperature to decline slowly and prepare to subsist on digestive absorption of stored fat. Some insects in larval, pupal, or adult stages may enter *diapause*—a daily or periodic routine of slowing down. The adaptation of birds to cold weather is, however, more varied; some stay active through the winter in a desperate search for food, and others migrate to warmer climates.

Light. Much has already been said to indicate the paramount significance of light in the biological world. To emphasize this point, one needs to refer to the all-encompassing role of photosynthesis in the continuity of the food-web and the cycling of materials through the biosphere (see p. 10).

The more subtle interactions between organisms and incident light may sometimes escape the detection of untrained eyes. We often take these interactions for granted. The green shoots of most plants turn toward the light. The sunflower is so sensitive to changes in the direction and intensity of light that its foliage and flower daily describe approximately an 180° angular shift in position from east to west in its orientation toward the sun. In contrast, the manzanita (bearberry) of the California chaparral orients its leaves vertically in order to offer a minimum surface to light.

Movement of plants in response to unilateral light is known as *phototropism*. In nature, this phenomenon occurs due to a greater accumulation of growth hormone on the less illuminated side of the plant organ. As a result, the darker side will grow faster than the lighted side, producing a curvature toward the light. Phototropism is intrinsically involved in the growth of vines, branches and leafy shoots that must arrange themselves for the maximum exposure to sunshine.

Adaptations to various light intensities can be equally vital. The relatively dark forest floor is favorable for some plants but inimical to others. Pine seedlings cannot grow in the shade of their towering parents, but maples and beeches can. The arctic flora efficiently utilize sunshine which is only a small fraction of that available in tropical areas.

The distribution of aquatic plants at various depths of water reflects their adaptation to low light intensities. While no vascular plants can be found at depths greater than 10 meters even in the clearest body of water, algae of various descriptions inhabit greater depths. For instance, Crater Lake, Oregon, is known to have algae growing at a depth of 120 meters where only 0.5% of sunshine can penetrate. Among all the algae, the red algae are exceptionally efficient in utilizing weak light.

There is still another dimension of incident light that is uniquely associated with the growth, development and distribution of plants. We are referring to the photoperiod or the daylength in a twenty-four-hour cycle. The phenomenon of *photoperiodism*—response of an organism to natural photoperiods—has added a fascinating chapter to our knowledge of modern biology.

In the year 1920, two scientists, W. W. Garner and H. A. Allard of the U.S. Department of Agriculture at Beltsville, Maryland, asked a basic question for the first time: Why do some plants flower in the summer, some in the fall, and still others in the early spring? The research that followed has unequivocally proved that most plants must receive several cycles of a *critical* amount of photoperiod before they can flower. On this basis, there are three types of plants to be found in nature: (a) *Short-day* plants that flower ordinarily in early spring or late fall (e.g. chrysanthemum, poinsettia, aster, goldenrod); (b) *long-day* plants that ordinarily flower in the summer (e.g. beet, clover, gladiolus, spinach); and finally (c) the *day-neutral* types that flower irrespective of the photoperiod (e.g. tomato, corn, snapdragon, cucumber).

An elaborate discussion of the mechanism of photoperiodic induction of flowering is beyond the scope of this discourse. It should be appreciated, however, that a flowering hormone is involved in this process. Defoliated plants (such as deciduous trees in winter) are not capable of responding to any photoperiod and, therefore, cannot produce flowering hormone. One may wonder, then, why redbuds burst out with profuse blossoms in the early spring before they show any sign of foliage. The answer lies in the fact that floral induction and flower-bud formation in these plants are completed before they enter a period of dormancy for the long winter. The warm spring days simply bring the flowers out.

It may be noted here that for the flowering of short-day plants, long uninterrupted nights are more important than short days. Plants such as chrysanthemum and poinsettia are conditioned artificially to unfavorable photoperiods to prevent their flowering before they are required for displays or commercial purposes. Then, a regular regime of short days (long nights) will readily make them blossom for the occasion. In the event of any inadvertent interruption of the long night—such as powerful street lights or the flashlight of a night watchman—the plants will fail to flower and continue to vegetate! This demonstrates the *all-or-none-type* stimulus which a critical photoperiod entails.

What can be a more dependable environmental cue than the photoperiod of a particular area? The earth's rotation around its own axis and around the sun make the photoperiod at a particular latitude on a particular day unerringly identical every year. This can hardly be said about temperature or other climatic factors prevailing in the same region. In the long process of natural selection, therefore, the elimination of the unfit from a mixed population of plants amounted to a simple test: Could the plants come to reproductive maturity (flowering) in synchrony with a desirable photoperiod? In other words, the requirement of a photo-sensitive plant for a particular daylength must correspond to the biogeographical area in which it grows and the season during which it comes into flowering. In this connection, plants growing above the latitude of

60° may be cited as examples. These plants must necessarily be either long-day types or day-neutral, since their brief growing season is marked by long photoperiods.

On the basis of the foregoing discussion, it would seem that most plants possess efficient time-measuring devices to detect favorable daylengths in which to flower. The precision of this time measurement can be utterly incredible even to an experienced scientist. Under experimental conditions, one variety of rice produces flowers 80 days from the day of planting, with a daily photoperiod of 11 hours and 50 minutes. With an artificial increase of daylengths by 5 minutes (11 hours and 55 minutes), the flowering of this plant is delayed by 20 days. With a further 5 minutes' increase in daylength, the flowering is delayed by an additional 20 days!

Light has its pervasive role to play in the animal world as well. The photoreceptor we call the "eye," has given most animals an extraordinary adaptive advantage which encompasses such behavior patterns as the response to warning, activities related to food-gathering, locomotion for safety and the display of color to attract mates or to confuse enemies.

The term *phototaxis* refers to a continuously adjusted movement of animals in response to light stimulus. Some insects are known to fly into fire and die self-inflicted deaths. Butterflies head for the sun to blind their pursuing predator. The bee-line formed by the busy bees is said to be in a precise angle in relation to the sun. The water scorpion swims in the direction of light. Even flatworms are known to move into a position determined by the light stimulus the two alert eyes receive. Animals change the direction of their locomotion as a result of unequal stimulation of otherwise symmetrical sense receptors. Of these, the eyes seem to be the most efficient. There are, however, animals without eyes—such as the mussel crab—which can respond to varying light intensities. *Mimicry* is still another form of adaptation and is most common among insects which, by virtue of resemblance with other species, often confuse their predators.

Vision is by far the most unique form of adaptation with regard to light. In addition to man and primates, many birds, reptiles and fish possess color vision. Curiously enough, most other mammals see only in different shades of gray. The bull charges the matador not because his cape is red, but because it is moving and, therefore, tantalizing! The longer wavelengths of the visible spectrum cannot be seen by insects. Instead, their vision overlaps an area of ultraviolet which man cannot see. What a bee sees in a bright red flower, for instance, is quite different from what the human eye sees.

With astonishing regularity, some birds from northern latitudes migrate every winter to warm southern climates thousands of miles away. With equally precise timing, they return north for nesting in the spring. What environmental cue enables these migratory birds to time their flights so efficiently? To emphasize what has been said earlier with respect to plants, "natural selection obviously favors the survival of those individuals whose response mechanisms use the most reliable source or combination of sources of information for the control of the annual reproductive cycle" (4, p. 137). Pioneer studies by a Canadian zoologist, William Rowan, and the investigations that followed have indicated that the natural photoperiod or daylength may be the single source of that information for migratory birds.

The instinct for northward flights during the early spring coincides with the growth of gonads or sex glands in the birds. What is more intriguing, the increasing daylengths in early spring trigger the hormonal reactions that lead to the growth of the testicle or ovary, which are otherwise inactive. The changing photoperiod primarily affects the sex-organ development, with migration as the secondary consequence. This has been further substantiated by studies of the reproductive cycles of non-migratory birds in relation to light.

The seasonal periodicity of light influences the reproductive cycles of other animals as well. Ferrets and deer appear to demonstrate reproductive readiness in synchrony with a critical photoperiod. Adaptive color development in fur and plumage (referred to earlier in this section), is known to be photoperiodically controlled. Hares and weasels are brown in summer and white in the winter. Finally, experimental studies have indicated that spawning of such fish as trout is triggered by the short days of December. The commercial fisheries can and do take advantage of this information, and stimulate spawning in the summer months by artifical control of daylength.

Dynamics of Adaptation. We have noted repeatedly in the foregoing sections that the story of adaptation begins with the interaction of an organism with its environment. This is, however, an incomplete picture unless we add to it the concept of multiplicity of organisms and the varied demands (in quality and quantity) they make on the environment and on one another. The situation becomes necessarily complex to the point that individual interacting factors are no longer discernible. Instead, all spatial and temporal aspects of the organism's life combine to assume a collective dimension—the *ecosystem*. For the sake of clarity, we shall discuss these aspects under three separate categories: (a) succession and climax; (b) endogenous rhythm; and (c) animal behavior.

Succession and Climax. The permanency of an ecosystem is more apparent than real. The aphorism "nothing is permanent except change" is appropriate to nature. An untended garden is literally taken over by the weeds which, in the course of time, will give way to other unwanted vegetation. The same is true about fallow agricultural land which, year after year, invites various species of grasses, herbaceous plants, and eventually seedlings of evergreen and hardwood trees. Concurrent with the vegetational changes apparent to the untrained eyes, the animal population also is modified.

The changes in the composition of a particular community over a period of time are far from chaotic and are, in fact, very orderly. The competitive sharing of environmental resources such as water, minerals, and light works to the detriment of some organisms and the benefit of others. Efficient utilization of limited raw materials will determine the survival, and eventual dominance, of one group of organisms over others. This systematic and inevitable change in a community is known as *succession*.

Primary succession begins in a bare area where no vegetation has previously grown. Newly exposed rocks, a newly formed island, or a small body of water are examples of such areas. In all instances of primary succession, the *pioneer* organisms initiate the process of change. In water, the plankton, microscopic algae and other unicellular organisms appear in great abundance as pioneers, and their death and decay set the stage for newcomers. The floating and rooted aquatic plants, as well as varieties of fish, follow. The nature of the habitat is transformed; prolific growth of floating and swimming organisms reduces the light intensity as well as the oxygen supply for the underwater vegetation. Added organic matter continues to build the bottom and reduce the depth of water. New organisms move in and partake of the new resources. Eventually, we see the highly familiar sight of rushes, reeds, cattails or wild rice encroaching into the water (Fig. 2–15). Among the animals, dragonflies, snails and frogs become increasingly frequent, as do birds and muskrats.

If the progression described above is left undisturbed, the change from aquatic to terrestrial forms of organisms will be complete. A new ecosystem may be born. It is not uncommon for an old-timer to observe these changes during his lifetime and tell a curious stranger that a lake was once situated where a pin oak or an alder is standing today.

The other form of primary succession begins on dry and barren rocks. The crustose lichens or mosses, by virtue of their capacity to resist drought and frost, are the pioneers in such formidable locations. These plants, by their gregarious growth habits, trap limited moisture and dust which together provide the necessary substrate for future species of plants to colonize. The speed of succession depends on how fast soil formation can take place. In tundra regions where this form of primary succession is common, the lower temperature slows down the process. As the depth of soil increases and more moisture is retained, hardy annual herbs and biennial and perennial grasses appear. Eventually, shrubs of the heath family (e.g. rhododendron) inhabit the area.

It may be noted at this point that various microcommunities formed by insects, fungi and bacteria illustrate the major concept of succession in reverse (the process of breaking down instead of building up). A fallen acorn supports a "tiny parade of life" as it awaits its complete degeneration (15, p. 120). The orderly entrance and exit of acorn weevil, fungi, moth and mites go unnoticed. In this case, "when organisms exploit an environment, their own life activities make the habitat unfavorable for their own survival, and instead create a favorable environment for different groups of organisms" (11, p. 141).

In contrast with primary succession, *secondary succession* results when a normal succession has been disrupted either by such natural hazards as fire and flood or by such artificial means as cultivation and lumbering. With man's intervention into nature, the

FIGURE 2-15

Primary succession in progress in a body of water. Photograph by David Morse.

outcome of secondary succession in some areas assumes a special significance. We shall return to this topic in Part III.

A golden opportunity to study secondary succession presented itself to the naturalists in the late nineteenth century. The central part of the volcanic island of Krakatoa, situated between Java and Sumatra, exploded, leaving a ragged 2500-foot peak surrounded by smoking pumice and ashes (14). Utterly desolate, sterilized, and essentially devoid of life, the island had to wait nearly half a century to regain its original lush vegetation and rich animal population. A fascinating story emerges as one follows the return journey of orchids, dandelions, coconuts, wild sugarcane and ferns as well as various birds and insects.

As a natural community matures, the number of species that are totally adapted to their habitat becomes fewer. Theoretically, then, a succession will eventually lead to a *climax* when all the inhabitants in the community are in balance with their environment, sharing the available resources (total production equaling total consumption). The dominant species become fewer in number but persistent in occurrence over a long period of time. A foreign species, a new disease or strange predator very seldom invades the community. We shall return later to the human implications of this phenomenon.

Endogenous Rhythym. It requires little scrutiny on our part to observe that most organisms periodically express certain patterns of behavior in synchrony with various environmental stimuli. These responses differ in timing and intensity. The chorus of birds signals the break of dawn; others take the cue from their noisy neighbors and begin their day's activity. With the sunset, bean plants lower their leaves. Nocturnal creatures such as owls, raccoons and moths make their appearances in response to darkness. The European shore crab scurries about the seashore precisely with the coming of high tide. A specific photoperiod stimulates a plant to bloom at a particular season of the year. The orbiting astronauts encounter several sequences of night and day within a single orbit but invariably prefer the work-sleep schedule to which they are accustomed on earth.

Biologists have long been intrigued by such rhythmic patterns of behavior as those mentioned above. The question remains, however: Is this rhythm endogenous (dictated from *within* the organism) or exogenous (caused by *external* stimulus)? In the case of photoperiodic induction of flowering, the stimulus is obviously external (see p. 62). Evidence weighs heavily, however, in favor of *endogenous rhythm* in most other instances. In other words, the behavior patterns in question, although synchronized with environmental cues, are not dependent on them. An internal *biological clock* tells time accurately to direct an organism in its complex interaction with the environment. This, of course, is a result of adaptation accomplished through innumerable generations.

"What time is it? Ask the quail or the bean seedling, the field mouse or the oyster.... They *know* what time it is." The high degree of assurance in the above statement (7, p. 18) results from the accumulation of startling observations. Let us examine several of these findings.

Bats display maximum activity in the dark, and the peak of this activity comes daily at the same time. What is more amazing is that this rhythmicity maintains its relative constancy even in continuous darkness.

The marine alga, *Gonyaulax*, luminesces most brilliantly in the middle of the night, and its cells divide most furiously in the early morning hours. Hundreds of miles away from the ocean and kept in the laboratory under constant dim light or darkness, this organism has been known to display periodicities in luminescence and cell division similar to those in its natural habitat.

The night-blooming jasmines open their flowers and emit powerful fragrance at night when the nocturnal flying insects pollinate the plants. Even under an artifical regime of continuous darkness or light, the same schedule of opening and closing of flowers persists (9, p. 378).

In one experiment (10, p. 361), bees were trained in Paris to eat in the early evening (French time) and flown overnight to New York. The strangeness of the surroundings did not deter the bees from appearing at the feeding place exactly 24 hours after the last feeding in Paris; it was then 3 P.M. in New York!

Similar experimental observations can be cited concerning the rhythms found, for instance, in mammalian body temperature, emergence of fruit flies from their pupal captivity, spore discharge in some fungi, or the color change in fiddler crabs. In all cases, it would appear that an internal timing device faithfully provides guidance to the organism even after the conditioning external signal has been withdrawn. This clock is like a wristwatch wound up at the birth of the organism and capable of telling time for extended periods without being reset.

There is another school of thought which seeks to qualify further the free-running nature of the biological clock. Professor Brown (2, p. 1537) contends that each organism is "plugged into" its environment, as it were, and therefore cannot escape environmental influence. Even in the experimental isolation of a light-proof box or under constant light, we may observe the remarkable sensitivity of certain organisms to such omnipresent and uncontrollable phenomena as earth's magnetic field, the rise and fall of the ionosphere or shifts in radio frequencies.

There are many kinds of rhythmic behavior observable in nature. The rhythm of an organism may be *circadian* (from the Latin *circa*—about, and *dies*—day) when it describes an approximately 24-hour cycle, as in the case of the leaf movement of bean plants. Rhythms can be *tidal,* as noted among green flatworms which come to the surface of the sand at high tide and bury themselves at low tide. The precision of the *lunar* cycle is matched only by some organisms that depend on it for reproductive efficiency. The best example is provided by the curious grunion fish that ride the crest of the highest tide to the California beaches and deposit their sperms and eggs. The fertilized eggs develop sufficiently above the tide mark during the next two weeks (one-half of a lunar month) until the next high tide washes the progeny safely into the open sea. Finally, there are the obvious *annual* or *seasonal* rhythms which seldom escape us as we observe wild flowers bloom every year or song birds signal their mating season.

The concept of endogenous rhythm becomes more intriguing as we consider behaviors that involve locomotion. The navigatory accuracy of some organisms staggers the imagination. How do the birds migrate unerringly over great distances year after year? How do Pacific salmon find their way back to the stream where they originally hatched? How is it that 3000 miles of open sea have seldom posed a problem for fur seals who must return to the Alaskan coast every mating season? How can one explain the behavior of green sea turtles negotiating vicious equatorial currents and countercurrents over a stretch of 1400 miles to find their nesting ground, Ascension Island—only seven miles long and six miles wide and inconspicuously located in the Atlantic Ocean?

Scientists do not have any reason to ascribe a sixth sense to these organisms! Instead, they have suspected—and some have obtained positive evidence for—celestial navigation systems implanted in the inherited instinct of these long-distance travelers. In other words, the biological clocks in these cases not only tell time, but also serve as compasses. The orientation may be with the sun, moon, stars or a combination of many terrestrial and extraterrestrial factors. Since the position of an organism shifts—continuously but predictably—in relation to these heavenly bodies, there must be a machinery with which ready corrections of their courses can be made. Experiments have confirmed this contention.

Our discussion of endogenous rhythm will be incomplete without a reference to that highly evolved mammal—man. It appears that human life is a composite of dependable and interrelated circadian rhythms. Disruption of sleep-wakefulness cycles following jet flights across several time zones constitutes an obvious example in this connection. In a highly industrialized society such as ours, interference with many functional cycles of the human body is commonplace. Professor Brown warns us: "If biological clocks are not reset to earth's rhythms, the organism will go berserk." Others, however, strike an optimistic note indicating that man's vulnerability in this respect can be much reduced by his rigid self-control which other organisms do not seem to possess.

Animal Behavior. In the foregoing section we have referred freely to such terms as "behavior" and "instinct" as they appear in the context of endogenous rhythm. We shall now turn to other forms of *animal behavior,* the study of which has become highly revealing and exciting in recent years. In fact, there are few facets of modern biology that can engage the interest of the uninitiated or the non-scientist more than observations in animal behavior.

Because of its central nervous system, the response of an animal to environmental stimuli is entirely different from that of a plant. Sensory perception of a stimulus is followed by motor impulses and, as a result, a behavior is manifested. The efficiency of sensory apparatus—sight, taste, hearing, touch and smell—varies greatly among animals, however. The eyesight of a zebra, for instance, is known to be considerably inferior to that of an ostrich, which, on the other hand, has poorer hearing than the zebra. Under the impending threat of an enemy, therefore, these two types of

FIGURE 2-16

Two rhesus monkeys are reciprocating friendly gestures. Photograph by the author.

animals will compensate for each other's deficiency as they graze in close proximity.

The stereotypes of certain animals and their behavior have been known to man for a long time. What we seldom realize, however, is that the cunning of a fox, the industriousness of an ant, or the wisdom of an owl reflects an extraordinary adaptation that resulted from natural selection. Intrinsic to an animal is its behavior which becomes increasingly complex as higher forms evolve. What could be more convincing evidence for this contention than the fact that now the animal kingdom even includes a species that is desperately trying to understand its own behavior!

In order for an animal to survive the pressure of natural selection, its inherited and learned behavior must provide it with three basic assurances: (a) the ability to evade or overcome a pursuing predator; (b) reasonable efficiency in finding food and shelter for itself; and finally, (c) proper display and utilization of its reproductive readiness.

A clever children's riddle goes something like this: "There are ten birds perched on a tree. A hunter shoots down one of them. How many are left?" An intelligent child does not have to be an expert in bird behavior to come up with the right answer—"None."

Like all other behavior, response to a potential danger, or the warning thereof, is innate among animals. In other words, the faculty of recognizing danger—through sight, sound or scent—is inherited by animals. The type of response varies substantially, however. A sand dollar buries itself in the sand to hide from a pursuing starfish; the hairs of the pronghorn antelope's white rump patch stand erect and, thereby, warn others of the approaching hunter; in the process of "playing dead" an opossum artfully makes itself utterly unattractive to an aggressive predator.

The nature of individual reactions to impending danger may assume a collective significance among the more social animals. The flocking and herding habits of some animals make it desirable from the point of view of survival that the entire group should respond to one signal in an identical manner. The devastating stampedes of buffaloes or elephants as well as the intense alarm calls of fleeing birds are obvious examples of such group reactions. A particular case in point is the behavior of starlings as they sight soaring falcons. The birds readily form a compact flock which presents a more difficult target for their attacker (Fig. 2–17). Similar

FIGURE 2-17

Starlings close rank as they spot the predatory falcon. After Tinbergen, The Study of Instinct *(Oxford: The Clarendon Press, 1949).*

cohesive effort can be noticed among harvester ants which recognize unauthorized strangers by specific scents.

Animals of the same species do not always form cooperative enterprises. They often fight their own kind in order to establish a place in their society—a process equally as vital as resisting or avoiding enemies. The most curious of such behavior is found in the "peck order" hierarchy of domestic chickens. The leader of the group, by its recognizable aggressiveness, wrests the right of initial access to food. In the event the leader is removed from the group, another leader—second in the hierarchy—takes its place.

Another form of social dominance can be observed in the territoriality of birds. With ceremonial calls and rituals directed against outside suitors, the males of the species "stake out" a piece of real estate and defend it as a breeding and feeding area. The promiscuous females, who often arrive later, wait until the termination of breeding to join the males in the defense of the territory.

Territoriality with distinctive behavior on the part of the males and females is common among the seals and certain fishes. In all instances this basic behavior appears to favor successful mating and safeguard against overtaxing the food supply, the level of which is critical during the breeding season (16).

Behavior related to food-gathering varies greatly among animals. Some hunt prey with various skills and regular frequency; others ingest the free-floating plankton of the ocean almost involuntarily; some swoop down on moving targets with unfailing accuracy, whereas scavengers survey from great altitudes animals that are already dead; some eat incessantly while still others store their food in neat packages against a rainy day.

One of the classical stories in connection with food-gathering relates to the "dancing of the bees" in search of nectar. The remarkable series of discoveries on this subject by Karl von Frisch (13) have intrigued biologists for several decades. A female scout detects the source of food and returns to the hive. She then arouses her hivemates with a "round dance" by first moving to her right and then to her left and repeating the sequence for several minutes (Fig. 2–18). The scent of nectar clinging to the body of the scout acts as an additional stimulant. Within a short time several hundred bees will be hovering over the food.

In the event that the food source is more than 50 to 100 meters away from the hive, the "round dance" is totally inadequate as a

Round dance

Tail-wagging dance

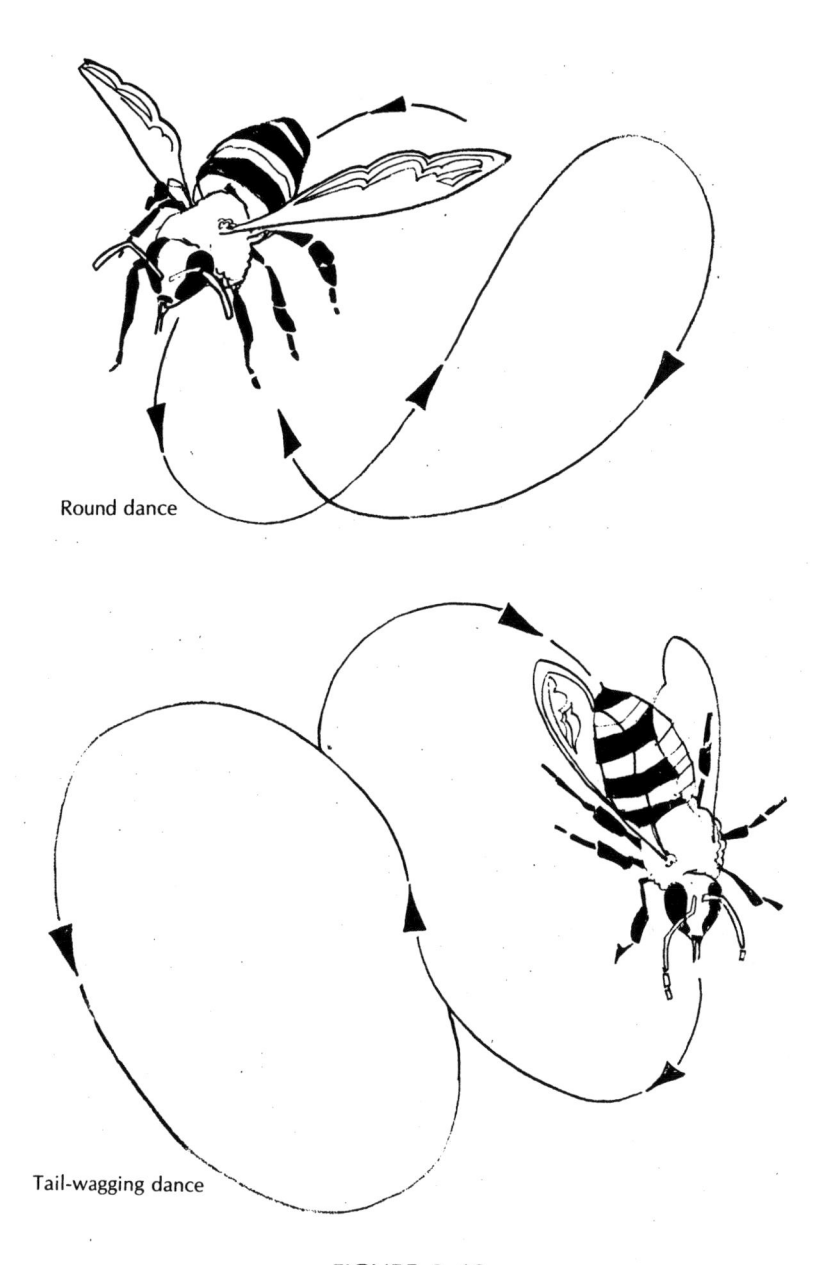

FIGURE 2-18

"Round" and "tail wagging" dance of the bee. Adapted from Karl von Frisch, Bees: Their Vision, Chemical Senses, and Language. Copyright 1950 by Cornell University. Used by permission of Cornell University Press.

directive. The scout then does a "tail-wagging dance" (Fig. 2–18). The vigor of the dance indicates the richness of the food source. As von Frisch puts it: ". . . the wagging dance transmits an exact description of the direction and distance of the goal. The amount and precision of the information far exceeds that carried by any other known communication system among animals other than man" (13, p. 3). The orientation of the wagging dance describes the direction of the food in relation to the position of the sun. The rapidity or sluggishness of the tail-wagging determines the distance between the hive and the destination. What is more incredible is that the scout bee never fails to compensate for the sun's shift in position as the time-consuming delivery of her message may sometimes warrant.

The instinct that accounts for an animal's seeking or building a durable shelter appears to be concomitant with its reproductive behavior. There are significant exceptions, however, as in the case of anthills, beehives, and the elaborately designed quarters of prairie dogs where the shelter reflects more the social organization of the animals in question than their immediate breeding plans.

The life cycle of a silkworm moth—especially the intricate construction of its cocoon—is a subject of unlimited curiosity. Following several moltings, the caterpillar stops feeding, anchors itself to the twig of a tree or shrub and builds itself a highly insulated winter home. The cocoon, which has a single exit on top, is composed of a thread of silk nearly a mile long and woven into three distinct layers. The compact inner and outer layers are separated by a middle, loosely-packed area. In the warmth and isolation of the cocoon, the caterpillar undergoes transformation until a moth is born and flies out of the encasement.

The enigma of the silkworm's behavior has been approached scientifically. Professor Van der Kloot (12) has attempted to unravel the interactions between gravity and the endocrine glands and tiny brain of the caterpillar. By his own admission, however, Professor Van der Kloot managed to ask more questions than he answered.

In keeping with the subject of the home-building instinct of animals, an observation from the author's experience in the tropics may be pertinent. Silhouetted against the sky, the pendulous clusters of nests built by crested cassiques on the fronds of towering palm trees swing precariously to and fro under the influence of violent gales. Yet these nests, made of carefully picked plant fibers and lined with leaves, never cease to prove their extraordinary durability. To withstand rain and storm during the breeding season, nature's test of survival obviously favored the instinct that could build such dependable shelters.

The continued existence of a species depends exclusively on its reproductive efficiency. In other words, an organism's ability to leave viable progeny constitutes a supreme test for the survival of its kind. We have discussed earlier how the endogenous rhythms of birds, fishes and turtles favor their reproductive success. We shall now turn briefly to those behaviors that account for the selection of mating partners, breeding, and postnatal care of the young.

The reproductive behavior of birds has been widely studied. Spectacular courtship rituals are common among such birds as grouse, prairie chicken, mallard, cormorant and herring gull (Fig. 2–19). The courtship is often communal, and sexual relations are promiscuous. The male of the species, following a successful defense of his territory, begins to woo the females with dances, display of colorful wings and various other physical postures. Distinctive behavior patterns connected with nest-building, feeding the young, and even postnuptial courtship are found among some species.

Among the mammals, the female in heat must be recognized by the prospective mate. This is often accomplished by the odor that advertizes reproductive readiness. The security of internal fertilization, the presence of placenta, and adequate prenatal growth of the embryo reduces the vulnerability of the newborn. As a result, it would appear that in mammals there is progressive reduction in wasteful multiple births, a common feature among the lower animals.

The parental instinct of caring for the young is of significant survival value and is highly developed among the vertebrates. Birds feed their helpless hatchlings on demand. In order to distract a predator from their chicks, some birds are known to feign injury themselves. Fishes such as sticklebacks build nests not only to facilitate elaborate courtship but to protect the young.

The most bizarre example of prenatal care of the young is provided by the emperor penguin of the antarctic. The female lays an egg and leaves it in the custody of the male partner while she goes off to feed. The solitary egg is hatched in the warmth provided

FIGURE 2-19

Courtship postures of prairie chickens. Photo courtesy of the State of Michigan, Department of Natural Resources.

by the abdominal folds of the male who faithfully performs this duty in a state of "incubatory fast." Having gorged on seaweed and navigated through great expanses of ice and snow, the female finally returns to the breeding ground and feeds the hatchling by regurgitation. The emaciated father, relieved of his duty, departs to feed himself!

The maternal behavior of eating the placenta and licking the newborn is common among many, though not all, mammals. The intensity of this instinct can be best exemplified by guinea pigs who, although habitually herbivorous, unhesitatingly consume their afterbirth (placental debris).

What we have discussed so far would tend to indicate that the primary drives for food, shelter and reproduction account for all the behavior of an animal and that these drives can be conveniently categorized under one term—instinct. It must be noted, however, that the term *instinct,* without proper qualification, is relatively meaningless. Does instinct describe only what an animal actually *does* or may it include also what it *can* do?

Scientists have come to believe that instinct includes the inherited ability to manifest a new behavior in a situation entirely alien to the organism. This potential, however, must be released by a minimum stimulus from outside. For instance, the egg-laying instinct of a female pigeon can be suppressed forever without the visual stimulus of a male. Birds will not feed their young except on demand marked by gaping mouths and raucous cries. A stuffed toy with the features of an owl will scare a bird, but will fail to do so if these features lack *minimum* tangible details. A male stickleback can be made to "copulate" with a cigar-shaped object only if this object possesses a swollen belly!

No biologist would question that the organic basis for a particular behavior pattern of an animal lies in its central nervous system which interacts with environmental stimuli. Nervous perception leads to hormonal secretion which, in turn, precipitates muscular and locomotory behavior. It is, however, exceedingly difficult, if not impossible, to trace this relationship experimentally when the behavior is highly complex. In such cases, only empirical observations are possible. To illustrate such complex behavior, one only needs to turn to the well known "suicide march" of the lemmings. These mouselike animals live in the mountains of Norway. As the density of their population periodically surpasses a certain limit, many leave their home and proceed toward the ocean, unrestrained by physical hazards. Many of them die on the way, and those that survive the trip end their lives by compulsively plunging into the ocean. The fraction of the population that avoided this ominous journey propagates a new generation of lemmings in the home territory.

Behavior patterns among animals present grades of complexity which seem to reach a peak among higher mammals— especially man—who can behave with reason as well as instinct. Although human behavior as a subject is not within the limits of our discussion, it is essential to remember that in the inexorable path of evolution, the process of human adaptation is no exception. The result is indeed remarkable.

This, then, is *Nature*. It is cruel, accommodating, and beautiful. It is an assembly of living and non-living, feigning apparent stability in time and space only to submit to their own replacement. From Himalayan peak to ocean bottom, from frozen Arctic and Antarctic to brooding tropics, life appears and disappears incessantly. Here, eons of time are imprinted on the blades of grass, on the wings of birds, in the movements of aquatic creatures, and in the language of the bees. Here poets, musicians and artists derive motivation. Here man has found an ecological niche for himself.

REFERENCES

1. Amen, R. D., "The Seed—an Unsolved Problem in Life," *Bioscience 14* (1964).

2. Brown, F. A., "Living Clocks," *Science 130* (1959).

3. Eisely, Loren, *Immense Journey*. New York: Random House, Inc., 1957.

4. Farner, D. S., "The Photoperiodic Control of Reproductive Cycles in Birds," *American Scientist 52* (1964).

5. Gates, David M., "Energy, Plants and Ecology," *Ecology 46* (1965).

6. _____, "Heat Transfer in Plants," *Scientific American* 213:10 (December, 1965).

7. Hicks, Clifford B., "What Time Is It?" *National Wildlife* 6 (August, 1968).

8. Licht, P., W. R. Dawson and V. H. Shoemaker, "Heat Resistance of Some Australian Lizards," *Copeio* (1966).

9. Overland, L., "Endogenous Rhythms in Opening and Odor of Flower of *Cestrum Nocturnum*," *Amer. J. Bot. 47* (1960).

10. Renner, M., "The Contribution of the Honeybee to the Study of Time-Sense and Astronomical Orientation," *Cold Spring Harbor Symp. Quant. Biol. 25* (1960).

11. Smith, Robert L., *Ecology and Field Biology*. New York: Harper & Row, Publishers, 1966.

12. Van der Kloot, William G., *Behavior*. New York: Holt, Rinehart & Winston, Inc., 1968.

13. von Frisch, Karl, "Dialects in the Language of the Bees," *Scientific American* Reprint #130 (August, 1962).

14. Went, F. W., "The Plants of Krakatoa," *Plant Life*. New York: Simon and Schuster (Scientific American book), 1957.

15. Winston, F. W., "The Acorn Microsere with Special Reference to Arthropods," *Ecology 37* (1956).

16. Wynne-Edwards, V. C., "Population Control in Animals," *Scientific American* 211:10 (August, 1964).

SUPPLEMENTARY READINGS

Billings, W. D., *Plants, Man and the Ecosystem* (2nd ed.). Belmont, Cal.: Wadsworth Publishing Co., Inc., 1970.

Brown, F. A., J. W. Hastings and J. D. Palmer, *The Biological Clock*. New York: Academic Press, Inc., 1970.

Daubenmire, Rexford F., *Plants and Environment* (2nd ed.). New York: John Wiley & Sons, Inc., 1959.

Etkin, William, and Daniel G. Freedman, *Social Behavior from Fish to Man*. Chicago: University of Chicago Press, 1969.

Moment, Gairdner B. (ed.), *Frontiers of Modern Biology*. Boston: Houghton Mifflin Co., 1962.

Sweeney, B. M., *Rhythmic Phenomena in Plants*. New York: Academic Press, Inc., 1969.

Wallace, Bruce and Adrian M. Srb. *Adaptation* (2nd ed.). Englewood Cliffs, N.J.: Prentice-Hall, Inc., 1964.

Went, F. W., "The Ecology of Desert Plants," *Scientific American* (April, 1955).

Part III

MAN

"It is not possible to pollute and ravage nature without deep agony and terror registering in the collective psyche of modern man."

Howard Thurman

The Dead and the Deadly. (Photograph by David A. Batchelder)

MAN

In spite of the obvious fact that he is an animal and is, therefore, an integral part of nature's biotic component, references to Man in the preceding chapter have been limited. The isolation of man from the total scheme of nature has been deliberate, not to accentuate the unfortunate dichotomy that already exists between man and nature in today's world, but to recognize it.

Man is indeed an animal—a highly developed mammal who, during the course of evolution, has come to occupy a uniquely dominant place in the biosphere. The question of man's dominance and uniqueness answers itself as soon as it is asked. Furthermore, there is no cosmic contest as to the validity of the question since *only* man has asked it. We shall return later to this intriguing philosophical controversy.

MAMMALS AND MAN

Biologically speaking, mammals constitute a supreme class of vertebrates that have evolved into abundant varieties over the last 70 million years (since the beginning of the Coenozoic era). Although mammalian forms such as rodents, carnivores, bats, and primates differ substantially in their outward appearance, they have a number of basic characteristics in common. All mammals are warm-blooded (homeothermic) and possess highly efficient circulatory and respiratory systems aided by compartmentalized hearts and the presence of diaphragms that separate the thorax from the abdomen. In contrast with its reptilian ancestors, a mammal's jaw is composed of only one bone that has supported progressively complex teeth. With the exception of the duckbilled platypus and spiny anteaters which lay eggs (see p. 42), all mammals are livebearers and nurse their young with milk from mammary glands.

Thought to have sprung from "an arboreal stock of small shrewlike insectivore," the primates constitute a distinctive group of organisms among the mammals. They are primarily characterized by: (a) extraordinary ability for locomotion facilitated by prominent elbow and shoulder joints; (b) enhanced mobility of digits on hands and feet; (c) development of binocular vision; and, most importantly, (d) increasingly large brain size compared to other mammals.

Among the primates there are two recognized groups in existence: (1) the prosimians ("pre-monkey") illustrated by lemurs in Madagascar and tarsiers that inhabit some parts of the Philippine Islands; and (2) *anthropoids* that include the new- and old-world monkeys, as well as the great apes and man.

The entire saga of the evolution of primates, it would seem, is one of independent lines of specialization to cope with arboreal life, ground-dwelling, posture and use of hands (Fig. 3-1). According to E.L. Simons of Yale University, the ancestors of four surviving apes—gibbon, orangutan, chimpanzee and gorilla—differentiated from monkey types during the oligocene epoch (30–40 million years ago). Chimpanzees and gorillas (especially the latter) abandoned the tree-tops, became distinct ground-dwellers and favored upright—though somewhat clumsy—walking postures. Their hands became increasingly free to grasp and use "tools." The persecution of Charles Darwin and, much later, the trial of the school teacher, John T. Scopes, can be traced to these momentous observations related to the curious resemblences between the great apes and man. Today most biologists are convinced by abundant fossil evidence that man *does* share a common ancestry with the apes, although his evolution digressed long ago from the lineage that led the apes into their present forms. This relationship may be simply expressed in the following diagram:

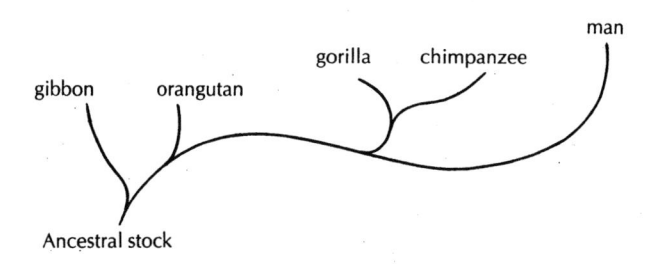

Recent biochemical studies of blood proteins have rendered further evidence that man's genealogical affinity with the chimpanzee or gorilla is substantially stronger than that which exists between any one of them and the gibbon or orangutan.

FIGURE 3-1

Phylogeny of the primates

FIGURE 3-2

The chimpanzee family. Phylogenetically speaking, it is more than a coincidence that the expression of love in this chimpanzee family is reminiscent of human experience. Photograph by Baron Hugo von Lawick, courtesy National Geographic Society.

ANCESTRY OF MODERN MAN

The archeological study of human ancestry is as frustrating as it is intriguing. Capricious nature, it would seem, has strewn the fossils of human skulls, jaw bones and teeth as a child leaves his toys at random, without concealing the trail of his activities. The essence of extinct life forms is preserved in geological deposits that punctuate the unfolding story of life. These relics of time, to be sure, often present the confusion of a jig-saw puzzle rather than a pattern of undisputed chronology.

Experts are divided in their opinions as to when and where primitive "ape-man" ceased to dominate and "modern man" emerged. It is reasonably certain, however, that this transition highlighted the last one million years of the pleistocene epoch when a series of glaciations covered large areas of the northern hemisphere with ice. As Bates emphasizes (2, p. 28): ". . . a considerable variety of man-like animals lived at different times and places in the pleistocene. One of them—the kind we know today—eventually won out."

Win out he did. Man's highly developed brain (an average of 1200–1500 cubic centimeters), as well as his sophistication in tool-making, made the crucial difference. It may be noted here, however, that modern man's brain is the result, rather than the cause, of his increasing efficiency in tool-making. Natural selection, over many generations of human existence, obviously favored those individuals that, by virtue of relatively greater neural and muscular coordination, produced better tools for hunting, building shelters, or protection. This progressive selection eventually led to a population which was characterized by immensely improved coordination —a quality that can stem only from highly developed brains. An educated, though not undisputed, guess indicates that modern man did not acquire the present attributes of his brain until approximately 50,000 years ago.

With brain and tools in his possession, it is impossible for modern man not to ask where he came from. Chronologically, the earliest success in this inquiry came in 1856 when human fossils were discovered in the German valley of Neanderthal. Subsequent examination of these fossils revealed that they belonged to men who lived in Europe and the Near East as recently as 100,000 years ago and were very similar to, but not identical with, today's man—*Homo sapiens*.

Presumably large-boned and short-statured, the "Neanderthal Man" possessed a brain nearly as large as that of modern man, made fine tools, and was capable of abstract thinking. Some anthropologists believe that a parallel stock of *Homo sapiens* from a different part of the world eventually crossed the path of "Neanderthal Man" and, as a result, either absorbed or annihilated the latter group.

Where did the "Neanderthal Man" come from? The logical answer to this question emerged from the discovery of "Java Man" in 1891 and "Peking Man" in 1920 in the form of middle pleistocene fossils (approximately one million years old). These fossils of jaws and skulls testified to the existence of bipedal upright individuals who constitute the intermediate forms of man, presumably of a different species—*Homo erectus*. The brain size (about 1000 cc) of these men was apparently large enough to qualify them as aggressive and successful predators. It later became evident from the abundance of *Homo erectus* fossils in the Old World, that these humans and their relatives dominated the middle pleistocene epoch.

In this backward journey in time, we have now come to the point when the search for human ancestors is ostensibly a search for the missing links between ape and man. From the quiet depths of African soil—often referred to as the cradle of human evolution—fossils of skulls, jaw bones and teeth, presumably belonging to the early pleistocene "ape-man," have been uncovered. In 1924 Raymond Dart, a South African professor of anatomy, ascribed a skull to the genus *Australopithecus* (from *Australis*, south, and *pithecus*, ape). In an effort to establish contact with the remote past, anthropologists have since flocked to Africa with the devotion of pilgrims. Many *Australopithecines*, as the "ape-men" are generally called, have been described from their fossils. Their average brain size is calculated to be 450–550 cc (compared to 350–450 cc of chimpanzees). Indications are that they were relatively short and bipedal with upright posture. The jaw bones were heavy and teeth exceedingly large. Evidence of their efficient hunting habits testifies to the existence of social groups among the *Australopithecines*. As Bates (2, p. 33) puts it succinctly: "If not men, then, they were at least well on the way to becoming men."

The sequence presented here describes an apparent lineage

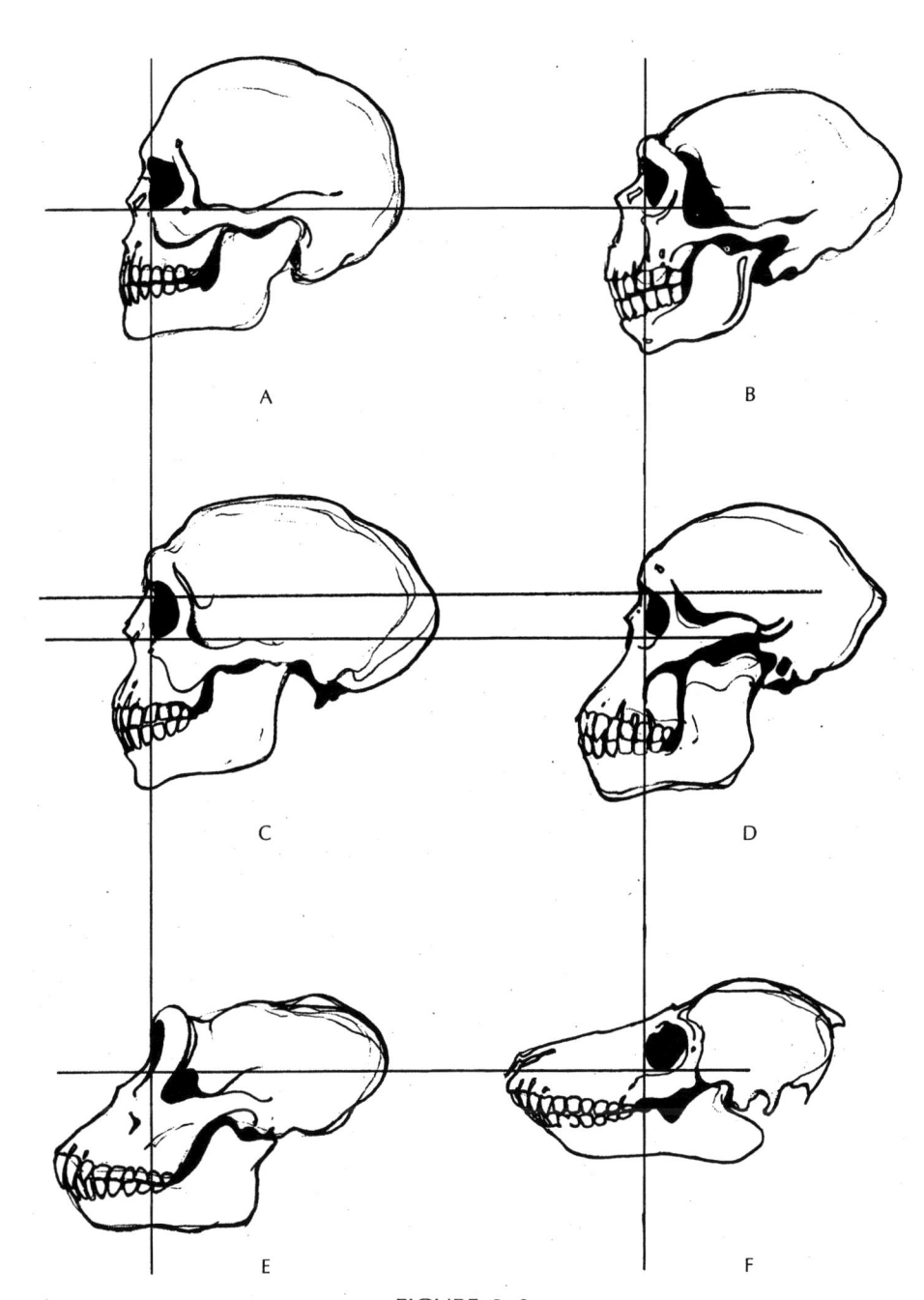

FIGURE 3-3

Progressive change in the size and orientation of various primate skulls. A: Modern man; B: Neanderthal; C: Homo erectus; D: Australopithecine; E: Chimpanzee; F: Lemur-type.

93

between the *Australopithecines, Homo erectus* and *Homo sapiens*. The validity of this cannot be examined without continued fossil evidence and highly sophisticated dating devices. In keeping with the context of this presentation, however, we are required to recognize only the *process* of evolution as it applies to *all* living beings. Since "life begets life" and "older order changeth, yielding a place to new," man cannot be insulated from this process. But could it be that by some fortuitous and random forces of selection, our ancestors with their increasing brain capacity added a new dimension to their own evolution?

MAN'S EVOLUTION: CULTURAL AND BIOLOGICAL

The evolution of the primate ceases to be entirely biological as soon as we call him human.

One late afternoon in July, 1969, mankind witnessed with awe and wonder the achievement of an incredible human feat—lunar exploration by man. Biologically, the heritage of these astronauts is shared by all *Homo sapiens* and can be traced to the ground-dwelling, tool-making Early Man. In other words, these intrepid space-voyagers do not constitute a distinct breed of individuals that exclusively propagate their kind, specially adapted for extraterrestrial living. They are only recipients of an extraordinary package of information assembled by many human brains and communicated to them with complete accuracy.

We have just alluded to the most significant development in man's evolution: his ability to gather and store information and to communicate it instantly to others. And what is more remarkable, an average human—barring congenital limitations—can use this information and thereby reap the harvest of several generations of "human progress" without having to inherit the good tidings himself from his parents.

To the best of our knowledge, no animal except man has yet acquired the advantage of directing the course of its own evolution. Man's biological evolution has opened a Pandora's box of what has come to be recognized as cultural or non-genetic psycho-social evolution.

All evolution is based on the fact that certain information is passed on from one group of individuals to others. A greater degree of variation in this body of information will increase the spectrum of variability among the recipients. In biological evolution, genes are the carriers of the information, and reproduction is the vehicle of transfer from one generation to another. The ingredient of cultural evolution, however, is not genetic in nature but "consists of ideas, inventions, traditions, laws, customs and all of the other learned responses by which society is regulated" (19, p. 174).

Cultural evolution is the work of the mind, and, therefore, uniquely human. Although man's biological evolution has not ceased—it never will—it is, however, dwarfed to relative insignificance beside the enormity of the fast pace of man's social and intellectual adaptations. Man's anatomical features, including brain size, have been subjected to little or no selective pressure during the last 40,000 years. Today we are justified in accepting the aphorism that, instead of adjusting to the environment, man adapts the environment to himself.

Man has built increasingly sophisticated tools and found ever faster means of transportation and communication. Elegant improvisations for underwater exploration compensate for natural limitations in the aquatic environment. Although we cannot run as fast as many animals, or fly like birds, we can make machines that transport us at supersonic speeds to distant lands, and even to other celestial bodies.

Paradoxically, man's apparent success in exploiting his natural environment has also produced unforeseen predicaments. In his world, the "unfit" do not necessarily die; but then, neither do the "fit" necessarily live.

MAN'S INTERACTIONS WITH NATURE

The foregoing discussion of man's cultural evolution, in effect, reflects man's interaction with nature. A closer examination of this interaction is in order at this point.

Historical Background. Historically, man has achieved ecological dominance by a series of long and diverse steps—some by chance and others by design. By virtue of gaining access to and advantage over his natural resources, he made the transition in his mode of life from food-gathering to food-producing and, finally, to food-processing. The highest premium, however, remained fixed on the sophistication of tool-making, be it for hunting, agricultural implements or industrial mass production.

The saga of man's interaction with nature is simply one of domestication of plants and animals for food and pleasure. With this drive came the need for intervention in the environment in order to reduce, if not eliminate, such natural enemies as parasites, large predators, pests and weeds.

Primitive men must have wandered from place to place and sought food like all other predatory animals. At times they found themselves in areas of relative abundance and "settled down" briefly to partake of the available berries, fruits and fish. Wild rice and assorted grains were no doubt consumed in this way. It is conceivable that careless scattering of residual seeds and grains around the manure-rich sites of temporary habitation produced the first revelation for agriculture (1). Men first eliminated the unwanted species of plants to protect the recognizably palatable ones. Later, the magic of cultivation of desirable plants came into existence.

There is considerable agreement among scientists that agriculture first attained full-scale operation about 10,000 years ago, although the technique of irrigation is of relatively more recent origin. There are also evidences that this new art of producing food may have sprung independently in different regions of the Old World such as Asia Minor and Southeast Asia. Many historical routes eventually spread the practice to the New World.

An intriguing argument has been advanced by Sauer (18, pp. 20–21): "Agriculture did not originate from a growing or chronic shortage of food . . . The saying that necessity is the mother of invention largely is not true. The needy and miserable societies are not inventive, for they lack the leisure for invention, experimentation and discussion." The thrust of this argument implies that settlements of human communities had to precede the collective undertaking of agricultural practice, the success of which demands expenditure of much physical energy.

The number of organisms that can share the resources of an area is limited by what biologists call the *environmental resistance*. This collective restriction consists of many components such as

FIGURE 3-4

A simple predator-prey relationship at the human level is represented by the Amazonian Indians fishing with bow and arrow. UNESCO Courier, 1954; Photograph by Monsieur D. Darbois.

food supply, predation and disease. For the primitive food gatherers who had to "eat on the run" as it were, an average area of two to ten square miles was needed to support a single individual. As a result, the "carrying capacity" of the entire earth during the Old Stone Age was about 20 million people. Needless to say, our ancestors did not fill that quota due to the high death rate.

With the advent of agrarian economy, man's relationship with nature changed drastically. The emphasis shifted to production. The added assurance for food was further fortified by efficient agricultural methods and larger family size which provided the much-needed human energy. The population ceiling rose accordingly to an estimated one billion (13, p. 3) although, in actuality, only 700 million people lived on this planet at the beginning of the Industrial Revolution in 1750 A.D. One should bear in mind that due to exclusive dependence on the human labor force and its natural vulnerability, the rise in the agrarian population stopped considerably short of being a runaway "explosion." This explains the apparent paradox that nearly 99 per cent of man's entire existence was spent to achieve what is less than one-fifth of our current world population of over 3.4 billion.

The Current Scene. What we are to consider in this section may come to some readers with the familiarity of a broken record. We refer to the cruel irony of modern man's industrial economy that plagues his waking hours with questions about the quality of life and even his future survival.

Since the Industrial Revolution, the exploitation of natural resources by man has been devastatingly complete. For his power man turned from the animate to the inanimate, from muscle to machine. Piecemeal efforts in production shifted to mass production and processing. Strong wind and flowing water generated electricity. Fossil fuels such as coal and oil were extracted from the heretofore undisturbed depths of the earth. The Atomic Age dawned with its ominous mixed blessing of growth and annihilation.

Population Explosion. With the enhanced ease of life, the chance of premature death was reduced as well. Life-saving medicines were invented, manufactured and made available. New insecticides killed vectors of lethal diseases. Rapid transport of food,

FIGURE 3-5

Labor and tool of agrarian life. A: Rice planting; B: Tilling the soil with the bullock plow. Photograph by T. D. Nag.

FIGURE 3-6

Tapping of a date palm, a natural source of highly-concentrated sugar sap. Photograph by T. D. Nag.

MAN

expansion of commerce, and better sanitation added to the remarkable success of death-control. An explosive combination of unrestricted human fecundity and high survival rate was triggered off.

At present, the world population is doubling in size once every 30–40 years with a *net* annual increase of approximately 70 million people. This means that at the current rate, the projected human population would be 7 billions by the year 2000 A.D. To put this awesome statistic in a more meaningful context, human beings as members of a single species (*Homo sapiens*), are outnumbered only by houseflies, codfish and sardines (2, p. 48)!

Until today, the increase in human population size with the concomitant increase in food supply was never deemed a problem, much less a crisis. After all, it was the basic urge for survival that motivated man in the first place to seek multiple ways of utilizing natural resources. What, then, is the cause of this transformation of a natural process to an unprecedented crisis?

It may be recalled (p. 10) that various organisms in the biosphere are interlocked in complex food-webs consisting of gigantic producer-consumer-reducer cycles. This complexity of interrelationships confers a natural resilience on the population size of an individual species. Biologists (4; 14; 21) repeatedly observed that the predator and prey populations are inversely correlated. As the predators grow in numbers, the prey population is reduced; a diminished availability of prey, in turn, brings down the population size of the predator. Therefore, barring any cataclysmic change in the ecosystem, the average population of a species remains relatively stable over a period of time. Some biologists attribute much of this stability to forms of social and reproductive behavior including responses to physical stress (crowding) and territoriality (p. 76).

The rate of human population growth is apparently out of line with the rest of the animal kingdom. Starvation or threat of it has failed to retard the reproductive efficiency of man; it has only increased the percentage of vegetating humans kept alive by modern medicine. Ironically, countries with greatly limited food resources have proportionately the largest population. With only 2.4 per cent of the total land area of the world, India is confronted with the task of supporting nearly 15 per cent of the world's population. The doubling time of the population of the United States, at the present rate, is about 63 years, whereas that of Costa Rica is 20 years. To shift to another continent, the corresponding figures for Japan and the Philippines are also 63 and 20 years, respectively.

The Malthusian prophecy of periodic famine and misery in the overpopulated countries has indeed come to pass. These catastrophes, however, have failed to act as a natural deterrent to population explosion. We are cautioned that we may still see Malthus fully validated by the middle of the 1970's (15).

Dilemma of Uneven Growth. A closer examination of the disparity between the functioning of human population and that of a natural population of wild animals and plants will unavoidably point to the socio-political institutions of man for a reasonable explanation. The Industrial Revolution began in England and spread through the rest of Europe and the Western hemisphere long before its impact was felt elsewhere. With some exceptions, today's Industrial man is still a Western man. The benefits of industrial society, such as drugs, insecticides and various labor-saving devices, however, found their way through the profit-motivated commerce and trade to the so-called underdeveloped countries. As a result, an additional imbalance was set up between the existing economy and the population. Agrarian birth-rates *artificially* combined with industrial death-rates. For instance, it took Sweden more than 80 years to bring her death rate to a stable low. The corresponding time was only 12 years in Ceylon, as the World Health Organization provided the necessary DDT to eradicate the cause of malaria, the principal killer. Many lands have been visited by the spectre of famine, although surplus food from the United States and Canada has managed to keep many hungry people alive.

There is nothing wrong with one person's hand feeding another person's mouth; in fact, it is indeed humane. However, there is no common collective goal in this haphazard economic scheme where one produces for profit and another consumes for survival. In reality, the socio-economic differences among the nations have never been more profound and disturbing. Oases of affluence have appeared in the vast desert of extreme poverty. With only 6 per cent of the world's population, the United States of America uses more than 50 per cent of the world's non-renewable resources. What is more stupefying, only 3 per cent of this country's *Gross National Product* is required to sustain the machinery of research and development on which much of this enormous wealth (nearly 45 per cent of the earth's total) entirely depends. The self-sustaining character of industrial mass production depends on this turnover rate. Considering the finite supply of raw materials in the earth's crust as well as other limitations, the prospect that such high productivity could occur everywhere at the same time is nil. The incompatibility between the world's population increase and its potential productivity has even pointed toward a possible reversal of industrial to agrarian economy (13, p. 7).

Too Little, Too Late? An analysis of the current scene will remain incomplete if we do not recognize the imaginative and creative efforts being made by individuals and organizations to alleviate the situation.

Understandably, our major battlefront has been in the area of accelerated food production and improving the nutrition in low-calorie diets. The gross food production throughout the world has increased considerably due to more accurate scientific know-how, efficient use of fertilizers, and new arable lands. Most recently, India recorded a harvest of 97 million tons of grain—an all-time high. The International Rice Research Institute in the Philippines has discovered "miracle rice" that has reportedly brightened the hope of eradicating hunger altogether from many parts of Asia. Under the auspices of the Rockefeller Foundation (17), scientists have bred a new variety of high-yielding, protein-rich corn—an ideal food for the millions of strict vegetarians in Asia.

Yet, in spite of these improvements, the *net* food production per capita has not been augmented; in fact, it has diminished in some countries as a result of expanding population. According to a conservative estimate, by the year 2000 A.D., the world food supply must triple to meet a reasonably adequate level of nutrition. At the suggested rate of one acre per capita, it would be an understatement to indicate that reclamation of new arable lands cannot keep pace with the rate of population increase.

Of equal concern is the problem of extreme malnutrition. This may appear to be less dramatic than death due to starvation, but increasingly it is proving to be the greatest stumbling block to the formative period of millions of children—the future citizens of the world. There seems to be little disagreement among scientists that a continuous protein-deficient diet can and does produce irreversible damage to the brain. Nearly two-thirds of the world faces this tragedy today.

The ingenious scientific approach of Professor William D. Gray of Southern Illinois University (8) is noteworthy in this connection. According to the basic law of a natural food chain, a tuna fish, in order to supply one pound of the canned product, has to consume 10 pounds of herring which, in turn, require 100 pounds of animals that must live on, proportionately, 1000 pounds of algae. The same energy budget is applicable to ranch and forage crops that pay us dividends in the form of beefsteaks. Mindful of the waste in the aforesaid energy transfer and the taxing economy that supports it in highly affluent societies, Professor Gray used waste substrate such as paper pulp in which certain species of fungi grew profusely. With the help of carbohydrate and added inexpensive nitrogen fertilizer, the fungi readily synthesized protein. The dry clear mycelium of the fungi was tasteless and odorless, and could be powdered and mixed with flour or other staples to increase the protein content from 2.5 to 15 per cent. Although the carnivorous Americans are not likely to stampede for the protein-rich flour as a substitute for steaks, the prospect of its use in some developing countries cannot be overemphasized.

In addition to Professor Gray's inventive approach, other efforts are being made to enrich the diet with protein. Fish-meals that have often been mixed with poultry feed are being further refined to improve human staples as well. Soybeans, originally cultivated as fodder, have proved to be highly nourishing as food substitutes or supplements such as yogurt, artificial bacon, and cold drinks. Oil refineries provide a potential substrate that submits to the biological actions of microorganisms to yield protein. Research at the University of California (Richmond) indicates that a pond (two-thirds of an acre in size) can produce sufficient algal protein to nourish 20 cows.

This brings us to the relatively unexplored area of oceanic flora and fauna as sources of food. Although some countries such as Japan have displayed imaginative efforts in the exploitation of fish and seaweeds, many local food taboos as well as bitter international controversies (20), have prevented the prospect of "farming the ocean" from being realized universally. Besides, the type of industrial machinery that is called upon to mass-produce food from the ocean would invariably pose a very serious, if not permanent, threat to the ocean itself. The scourge has already begun (7; 11).

THE BY-PRODUCTS OF AFFLUENCE

Depletion of natural resources, environmental pollution, and urban blight have been sarcastically, but not unjustifiably, described as the "gross national by-products" of a highly industrialized society. The "protective callousness of affluence" as Margaret Mead calls it, has transformed man's dream of harmonious interaction with nature into a nightmare of frightening self-defeat. This paradox of modern man has come into such sharp focus recently that its documentation is unnecessary. We shall only emphasize the salient features of this problem as exemplified by the current American scene.

Depletion of Natural Resources. It should be abundantly clear that organisms and the environment in which they live are not haphazardly related. Instead, they form orderly and integrated units of existence. These units are called *ecosystems*. An isolated lake with its flora and fauna forms a natural ecosystem where the input and output of energy, as well as the cycling of materials, are in a steady state. A well-maintained aquarium, on the other hand, can be an example of an artificial ecosystem.

What if the consumption of one type of raw materials by a group of organisms suddenly exceeds the established rate? What happens when the cycling of renewable resources is impeded by increased or retarded biological activity in the "chain of command," as it were? Obviously, the resources will be depleted beyond recovery and the dynamic equilibrium in the ecosystem will collapse. This is essentially the warning modern man has received repeatedly as he has turned for survival and well-being to such resources as land, forest, and water.

During the early conservation movement in the United States, small gains were noticeable in certain areas. These gains, however, have been rendered insignificant in the context of the current industrial and technological boom. Today, the depletion of natural resources is a direct, if not inevitable, result of an ever-expanding population which the society must support. Proven scientific tools and tested technology provided assurance that the country could meet the emerging demographic demands. This assurance does not appear to be justified, however. Lack of ecological knowledge failed to deter profit-motivated industries and other interest-groups who have for many years operated through effective political lobbying and investment of enormous capital. Responsibility has become increasingly diffused among people in high office. As a result, no collective goal has emerged. This phenomenon of "the left hand not knowing what the right hand is doing" can be illustrated by the following scenario based on a description by Aldo Leopold (10, p. 537):

> A road crew cuts a grade in a clay bank and permanently destroys a trout stream, while another crew upstream is trying to improve it by building dams and shelters. An outdoor recreation crew builds fireplaces and burns the trees and shrubbery felled by lumbermen who have very little understanding of game food resources. Finally, the open fields, so essential for such animals as deer and partridge, are planted with pines in the name of conservation!

Land is the ultimate substrate in which we build, grow, dream and play. It is many-faceted, providing services, products, and sanctuaries. It is the backbone and spirit of any nation.

The total forest land in the United States today amounts to about 509 million acres, of which 28 per cent is owned by the federal government and 13 per cent by industry. The remaining 59 per cent belongs to farmers and private owners who have no specific motivation or plan to conserve their land.

Croplands as well as forests have been subjected to intensive exploitation by farming and lumber enterprises. Arable land of 3.91 acres per capita in 1930 has dwindled to 2.6 acres per capita today. We are told that the danger point may be reached by 1975. The extent of our demands on forest products is similarly unprecedented — demands that by the year 2000 A.D. would be impossible to meet. A dramatic statistic from the 1949 yearbook of the United States Department of Agriculture may clarify the point further: In one year the Sunday issues of the *New York Times* amount to 125,000 tons of newsprint which, in turn, utilize for paper pulp the annual growth from 1200 square miles of Canadian forest land!

The question of forest, cropland or soil conservation today is intrinsically connected with the blistering pace of highway construction, and the demands for recreation centers and additional national parks. It involves the direct effect of soil erosion and loss of watershed. For every mile of a four-lane highway, approximately 50 acres of land must be "consumed" and rendered unproductive from the ecological point of view. What is more tragic, however, is that due to inadequate and financially expedient planning, stretches of highway must be rebuilt as superhighway, thereby adding further insult to the injury already suffered by the proximal land (16, pp. 56–7).

If land is our primary substrate, water is the cradle of our life (p. 17). There is no question about the total abundance of water contained in our streams, rivers, lakes and topsoil. The question is, how much of it is usable at present and will remain so in the foreseeable future — both for human consumption and crop irrigation?

Americans, who have the dubious reputation of being "water gluttons," use over 360 billion gallons of water a day. This consumption is projected to increase 66 per cent by 1980 and 150 per cent by the year 2000. The receding water-table in certain parts of the country has already alarmed the hydrologic engineers. This concern applies to the relatively wet eastern seaboard as well as areas where dams and long-distance conduits must direct water to homes and industrial plants. To appreciate the intensity of the problem, one merely has to turn to Southern California. Don Marquis' serio-comic utterance as expressed through his ingenious character "archie,"[1] the cockroach, appears as a prophecy at this point: "it wont be long till everything is a desert from the alleghenies to the rockies the deserts are coming the deserts are spreading the springs and streams are drying up one day the mississippi itself will be a bed of sand ants and scorpions and centipedes shall inherit the earth [sic]."

The term *conservation* has often been heard in the righteous indignation of the wide-eyed rugged naturalists as they seek to save a wilderness area or create one more national park. Just as the judicious use of resources in a business enterprise is construed to be good management, the conservation of national resources should also be a matter of management. What remains as the thorny and unanswered question is: who manages, and for whom?

---

[1] Don Marquis, *the lives and times of archie and mehitabel* (Garden City, N.Y.: Doubleday & Company, Inc., 1935), p. 15.

In the wilds, the balance sheet of the quality and quantity of growth—the input and output of energy—is drawn by such natural forces as climate, successional trends and migration of plants and animals. Here the same process that wounds also does the healing. There is no profit or loss by design. In communities where man is the major biotic factor, however, planned exploitation of resources invariably—though not unavoidably—entails a design for profit or loss. Due to lack of coordination and ecological knowledge, as mentioned earlier, this exploitation has often terminated in complete destruction.

In the true sense of the term, conservation implies neither austerity nor altruism. It only refers to a principle of applied ecology unabashedly designed for the welfare of man. It makes man's intervention in nature economically and ethically acceptable until it fails to deliver benefits to the people and becomes, instead, subject to the inanimate corporate power of "tainted" technology. There was nothing unethical about the pilgrims' felling trees and clearing the forest to build dwellings. However, with our present relatively sufficient knowledge about the denuded forest, loss of watershed and soil erosion, we have come to recognize indiscriminate lumbering as a moral issue.

The rationale for conservation of natural resources can be jeopardized in another way. We are of course referring to the outrageous numbers of people that our finite resources must support. "So profound and all-pervasive are the effects of exploding population and expanding technology that efforts to deal with environmental problems are unlikely to be effective in a long-term sense unless we turn our attention to these two root factors." This is a sobering pronouncement to be found in the *conservation yearbook #4* (1968) published by the United States Department of the Interior.

It is obvious that we cannot conserve our wilderness and have, at the same time, more and more people demanding space for their own existence. We cannot protect our forest and, at the same time, maximize our demand on forest products. Yet national parks, forests and lakes become our magnificent treasures and life-support systems—*not* in a vacuum but in the minds and hearts of people. A Douglas fir on the Cascade slope of Oregon or a blade of grass in the sand dune of New Mexico appears to transcend its purpose of existence or even define it in a symbiosis with other organisms—man included.

As population grows, so do the human demands for more wilderness, more recreation areas and more highways to get there. In view of this predicament, our conservation movement has failed before it begins. We need to launch another crash program, evidently with a more meaningful thrust: "Conservation of the *people*, by the *people* and for the *people*!"

Environmental Quality. The aphorism that "nothing is permanent except change" applies admirably to nature (p. 66). We have recognized that, barring such catastrophes as floods, fire or earthquakes, changes in a natural ecosystem follow a relatively predictable pattern. If, however, this change is accelerated by artificial means and in a haphazard fashion, the results are no longer predictable; we become finally aware of them as they appear before our eyes, often with a startling suddenness. In today's industrial society, the most significant thing about change is the speed with which it is occurring.

To cope with the needs of an expanding population and to maintain a semblance of "well-being" among the taxpayers, the technological machinery has violated man's unwritten contract with nature, abused power to generate more power and defiled our environment with mindless tampering. Society is in a technological trap which perceives progress as doing more and more of the same, but bigger and faster. It assumes that "tomorrow will be but a horrendous extension of today" (6, p. 2).

Depletion of natural resources is measurable in such terms as tons of ores, numbers of felled trees, or acres of vanishing wilderness. We do not, however, have concrete measures to determine the rate of decline in environmental quality—the physical and spiritual threshold of our very existence. Air, water and solid-waste pollution are only cumulative results. The causes of this deterioration were initiated long ago as man released his power on this planet and attempted to thwart the natural processes. Some may even argue that "civilization" and pollution of the environment must necessarily go hand in hand.

Pollution of the air is not a new phenomenon. The ancient Romans complained about the "heavy air of Rome." What is new, however, is the intensity of the problem and the sudden realization that there is an "air of finality" about the air we breathe.

The air in our cities and suburbs is deemed polluted since it contains various noxious gases and particles in addition to its normal components—oxygen, nitrogen, carbon dioxide and water vapor. This spells both short-term disaster and long-term public health hazard.

Automobile exhausts are primarily responsible for air pollution today, contributing, on the average, 65 per cent of the total pollutants. Treading faithfully on the heels of automobiles as polluters are our industrial and power plants, burning fossil fuels and spewing out large quantities of particulate and gaseous pollutants. Sulfur and nitrogen oxides as well as hydrocarbons and carbon monoxide form the bulk of our current annual aerial garbage of over 140 million tons. Polluted air is known to crack rubber, discolor paint, corrode stone statues and damage crops. What must it be doing to human bodies?

Our cities, where 70 per cent of the country's 200 million people live, have specific situations to cope with. These massive conglomerations of brick and concrete have macroclimates that make them relatively warmer, drier, dustier and foggier. In the event of *temperature inversion* which usually occurs in the autumn and winter months, the inhabitants have little alternative but to share stationary polluted air, at the normal rate of 2000 gallons per day per person.

Carbon dioxide in the atmosphere absorbs infrared from solar radiation and contributes to the total thermal energy. The green vegetation in rural areas continuously removes the carbon dioxide in photosynthesis. In cities, where industrial plants add considerable amounts of excess carbon dioxide to the air, the average temperature rises appreciably. Scientists believe that this temperature rise is a world-wide phenomenon—often referred to as the "greenhouse effect." The concern for the melting of polar ice and consequent rise in the level of the oceans is no longer the fabrication of an alarmist.

Most Americans today are well aware of the serious problem of water pollution. Here is still another example of biological warfare directed against ourselves in the pursuit of questionable gains.

As indicated earlier, aquatic ecosystems such as lakes and streams are vulnerable to pollution as a home aquarium is to casual maltreatment (over-feeding, poor aeration, etc.). Algae are obviously the producers in a body of water, while protozoa, insects, arthropods and finally, the fish form the hierarchy of consumers.

MAN

Upon the death of these organisms, decay is accomplished by the all-important reducers—bacteria. The recycling of materials continues.

The biochemical oxygen demand (BOD) plays a very important part in the maintenance of an aquatic ecosystem. Quantities of organic matter deposited in water depend on the population of dead organisms. If this accumulation becomes too great for the available oxygen to break down, a huge "underwater cesspool" will eventually result.

Although water pollution can be traced to homes and individuals, two major factors stand out as we examine this problem on a nationwide basis. Nitrate and phosphate fertilizers, which are used by the farmer in colossal quantities, find their way into adjoining lakes and streams by seepage and run-off and make the algae "bloom." The well-known "pea soup" appearance of Lake Erie in recent summers testifies to this "bloom." In the process of decomposition, large quantities of organic materials from luxuriant algal growth deplete the oxygen supply to dangerous levels.

Abnormal increase in the water temperature is another cause of water pollution. Electrical and industrial plants, invariably located near lakes and rivers, use water for cooling purposes and return it—hot and poisonous—to the reservoir. Higher temperature lowers the oxygen content of water, allows inordinate growth of algae and other plants, and eventually makes the habitat unfit for any living thing. Unless trends are reversed, no waterway can be protected from this thermal pollution, be it the Hudson River or Lake Michigan, the Rhine River or Lake Baikal.

The old adage, "Running water purifies itself," is biologically justified. The recuperative capacity of water, however, is not unlimited, considering the stress brought about by agricultural and urban run-off (3). If we add to this lesson our recent experience that "oil and water don't mix," the situation becomes ominous indeed. We recall with horror how the spillage of over 100,000 tons of oil from the American tanker, *Torrey Canyon,* devastated seabirds and shellfish off the coast of Cornwall in March, 1967. Since that time similar incidents have occurred elsewhere. Drillings off the Gulf coast and the coast of southern California continue to take their toll of marine organisms in estuaries and tidal marshes.

The message of Rachel Carson's *Silent Spring* (5) is finally coming through to us, loud and clear. Synthetic pesticides directed against the lowly enemies of farmers and fruit-growers have journeyed inexorably into our food additives, milk, fruit, eggs and meat. To dramatize this fact, Professor Daniel F. Jackson of the University of Syracuse once assured health conference delegates following a breakfast, "You can see, you are really getting more for the money than appeared on the menu" (quoted in 16, pp. 198–99).

Since its discovery in 1946, the annual production of DDT has increased from one million pounds to over one billion pounds. Due to its non-biodegradable nature, it stays adrift in our food web and becomes increasingly concentrated in the bodies of consumers. This process is known as *biological magnification* (Fig. 3–7). By the admission of the U.S. Department of Interior, DDT, even in extremely low concentrations, has proved debilitating, if not lethal, to most game birds and animals as well as fish and crabs.

Of special concern is the productivity of the phytoplankton in the ocean which carries its share of DDT from the enormous stockpiles distributed in all parts of the world. In view of the extreme sensitivity of algae to DDT, the world's oxygen supply—more than 80 per cent of which comes from the ocean—is in jeopardy.

Obviously, a major mistake can obscure actual gains. The importance of pest and disease control in relation to crop production and public health cannot be ignored. Yet one wonders if society must always wait this long to discover that biodegradable pesticides and low-phosphate detergents can serve our needs admirably without polluting our environment.

Perhaps no other problem points more directly and visibly to our population growth and total dependence on technology than solid-waste pollution. In fact, the amount of garbage and rubbish is increasing twice as fast as the population. The cost of solid-waste disposal in America today is exceeded only by that of highways and schools. Every eight seconds another American is born. This newcomer, as Rienow and Rienow (16) point out so graphically, demands and partakes of the industrial "bonanza" for the next seventy years. In his consumptive role of opulence he contributes to the staggering amount of solid waste. According to the United Nations Economic and Social Council (May, 1969), the annual solid-waste products in the U.S.A. include 20 million tons of paper, 7 million automobiles, 48 billion cans and 26 billion bottles.

Sanitary landfills, which can literally "make a mountain out of a molehill," have filled many abandoned quarries and lowlands,

FIGURE 3-7

Biological magnification of DDT as the much-abused pesticide passes through the food chain. Modified by permission of Time-Life, Inc., from Time magazine, 1969.

returning dividends to some communities in the form of parks and playgrounds. Composting and mulching of rubbish have also provided effective organic fertilizer. Unfortunately, however, such "immortals" as aluminum cans and plastic wares, which form a substantial bulk of the by-products of our modern living, cannot be recycled; they just accumulate in a one-way traffic as the "time capsules" of man's industrial affluence!

Urban Blight. If environmental pollution is considered a pathological result of our style of life in a highly industrialized society, then *urban blight* must be the terminal disease of that same pathogen. The only difference lies in the manifestation of the symptoms. In addition to depicting a precarious outward pollution, urban blight testifies to symptoms of man's inner environment—the bankruptcy of his spirit, if you will. In a scale illustrating the relationship of an organism to its environment, urban blight is the lowest limit that the human animal can reach in direct response to an *unnatural* environment. Phenologically, it is evolution in reverse!

In ecological terms, we have recognized that anything natural is neither good nor bad; it is *neutral,* awaiting the test of time and space. Anything simple, such as the multitude of unicellular algae in the ocean, is *vulnerable* to unrestricted and defenseless death. Desperate and prolific reproduction is the only means for the survival of its kind. Any system that is complex in its interdependence, however, is *efficient* and *resilient.*

Consider for a moment that human society is the epitome of a complex system. If the network of this interdependence is severed at various points, both its efficiency and resilience are lost. Urban blight is the extreme example of this crisis in human ecology. Here, too, the reproductive rate becomes a critical factor. For the same ecological reason that rapid reproduction in algae is an indispensible adaptation, human overpopulation has become a total threat to man's survival. "It is the top of the ninth inning. Man, always a threat at the plate, has been hitting Nature hard. It is important to remember, however, that Nature bats last!"[2] Urban blight points to the fact that Nature did bat last and, what is more, came from behind to win.

---

[2] Cited by Paul Ehrlich in *Eco-catastrophe* (San Francisco: Canfield Press, 1970), p. 14.

It has been said that the natural resource least used by man is man himself. Nowhere is this more evident than in the urban ghetto —a man-made cesspool, not unlike the bottom of Lake Erie. The ecological disruption here, however, appears as a "city syndrome" with socio-economic and political manifestations. In this respect, the life of a city can be equated to the life of an organism where each member must relate to the others to avoid decay and progressive loss of identity. If we are to appreciate Lewis Mumford's identification of a city as a "process," the ecological dislocation of this process will indeed explain what Hall (9, p. 155) has termed a chain reaction, "creating a series of destructive behavioral sinks more lethal than the hydrogen bomb."

This, then, is our discourse on Man. An object of nature, he lives, grows and dies. The predominant species of the biosphere, he designs, commands and exploits. His adaptations have passed the "fitness" test of the environment admirably. The paradoxes of his present style of life, however, make it evident that he is failing the test of time. The nomadic hunter, the toiling farmer and finally, the mechanized industrialist, man has come to a peculiar juncture of his existence. He must either seek fulfillment in isolation or reinstate his partnership with Nature. Does he have a choice?

REFERENCES

1. Baker, Herbert G., *Plants and Civilization* (2nd ed.). Belmont, Cal.: Wadsworth Publishing Co., Inc., 1970.

2. Bates, Marston, *Man in Nature* (2nd ed.). Englewood Cliffs, N.J.: Prentice-Hall, Inc., 1964.

3. Benarde, M. A., *Our Precarious Habitat*. New York: W. W. Norton and Co., Inc., 1970.

4. Calhoun, John B., "Population Density and Social Pathology," *Scientific American* 206:32 (February, 1962).

5. Carson, Rachel, *Silent Spring*. Boston: Houghton Mifflin Company, 1962.

6. Dubos, Rene Jules, "The Human Environment in Technological Societies," *Rockefeller University Review* (July–August, 1968).

7. Ehrlich, Paul R., *The Population Bomb*. New York: Ballantine Books, Inc., 1968.

8. Gray, William D., "Fungi as a Potential Source of Edible Protein," *Activities Report 17*, Research and Development Associates (1965).

9. Hall, Edward T., *The Hidden Dimension*. Garden City, N.Y.: Doubleday & Company, Inc., 1966.

10. Leopold, A., "Conservation Economics," *Jour. Forest.* 32 (1934).

11. Marx, Wesley, *The Frail Ocean*. New York: Ballantine Books, Inc., 1969.

12. McAlester, A. Lee, *The History of Life*. Englewood Cliffs, N.J.: Prentice-Hall, Inc., 1968.

13. Neil, R. E., "The Population Explosion in Historical Perspective," *Bates College Bulletin* (special), Lewiston, Maine, 1967.

14. Odum, Eugene P. and Howard T. Odum, *Fundamentals of Ecology* (2nd ed.). Philadelphia: W. B. Saunders Company, 1959.

15. Paddock, William and Paul Paddock, *Famine—1975: America's Decision, Who Will Survive*. Boston: Little, Brown and Company, 1967.

16. Rienow, Robert and Leona T. Rienow, *Moment in the Sun*. New York: Ballantine Books, Inc., 1967.

17. *Rockefeller Foundation Quarterly #1*. Rockefeller Foundation Special Report, *A Partnership to Improve Food Production in India*, 1969.

18. Sauer, Carl O., *Agricultural Origins and Dispersals*. New York: The American Geographical Society, 1952.

19. Stebbins, G. Ledyard, *Processes of Organic Evolution*. Englewood Cliffs, N.J.: Prentice-Hall, Inc., 1966.

20. Wooster, Warren S., "The Ocean and Man," *Scientific American* 221:50 (September, 1969).

21. Wynne-Edwards, V. C., "Self-Regulating Systems in Populations of Animals," *Science 147* (March 26, 1965).

SUPPLEMENTARY READINGS

Carter, L. J., "World Food Supply: Problems and Prospects," *Science 155* (1967).

Chalquest, R. R., "Feeding the World," *Bioscience 16* (1966).

Clark, John R., "Thermal Pollution and Aquatic Life," *Scientific American* 220:14 (March, 1969).

Cole, L. C., "Man's Ecosystem," *Bioscience 16* (1966).

Commoner, Barry, *Science and Survival*. New York: The Viking Press, Inc., 1966.

Cox, George W. (ed.), *Readings in Conservation Ecology*. New York: Appleton-Century-Crofts, 1969.

Crow, J. F., "The Quality of People: Human Evolutionary Changes," *Bioscience 16* (1966).

Dasmann, Raymond F., *Environmental Conservation* (2nd ed.). New York: John Wiley & Sons, Inc., 1968.

Davis, Kingsley, "Population," *Scientific American* 209:24 (September, 1963).

Edwards, Clive A., "Soil Pollutants and Soil Animals," *Scientific American* 220:12 (April, 1969).

Harrison, Richard J. and William Montagna, *Man*. New York: Appleton-Century-Crofts, 1969.

Helfrich, H. W., Jr., *The Environmental Crisis*. New Haven, Conn.: Yale University Press, 1970.

Novick, Sheldon, *The Careless Atom*. Boston: Houghton Mifflin Co., 1969.

Population Bulletin 22. Population Reference Bureau, Inc., (1966).

Shepard, Paul and Daniel McKinley (eds.), *The Subversive Science: Essays Toward an Ecology of Man*. Boston: Houghton Mifflin Company, 1969.

Whiteside, T., *Defoliation*. New York: Ballantine Books, Inc., 1970.

Williamson, F. S. L., "Population Pollution," *Bioscience 19* (1969).

Woodwell, George M., "The Ecological Effects of Radiation," *Scientific American* 208:28 (June, 1963).

―――, "Toxic Substances and Ecological Cycles," *Scientific American* 216:19 (March, 1967).

Young, L. B. (ed.), *Evolution of Man*. New York: Oxford University Press, 1970.

Part IV

PROSPECT

"Ah, but a man's reach must exceed his grasp."

Robert Browning

It is neither "black" nor "white;" it is both, reciprocating in a perceptual parallax.

PROSPECT

In the foregoing sections, attempts have been made to underscore a skeleton of abiding commonality among all living organisms. From algae to flowering plants and from protozoa to people, there exists an organic, pervasive continuity grounded in time and space. Obviously, the process of biological evolution which embodies heredity (unity) and variation (diversity) provides us with the scientific model which explains this pervasiveness.

And yet, "the products of man's creative urge reflect the notion that everything is referred to himself in whom and through whom all things are integrated and justified."[1] In other words, reality is anthropocentric. The *unity in diversity* in question does not exist by itself; it unfolds as a process of rational thought. Nothing about Nature can be conceptualized without direct or vicarious human intervention. No wonder, then, that a discourse on man and nature should invariably gravitate toward man himself—the ambivalent, rational animal who depends on bread to live but "does not live by bread alone."

In order to characterize the gaping discrepancy between man and nature, a conceptual scheme must be sought. Extending beyond the traditional jurisdiction of a scientist, this scheme ideally should allow us to cross freely the arbitrary lines of demarcation between the sanction of facts and the command of feelings, between reason and articles of faith.

ALONE IN THE CROWD?

Professor Michael Polanyi, the famous physical chemist at Oxford University, has warned (2, p. 152), "The scientific ideal of an absolute truth divorced from human judgment is a dangerous fallacy that seriously impedes scientific progress." The contemporary

---

[1] Richard J. Harrison and William Montagna, *Man* (New York: Appleton-Century-Crofts, 1969), p. 366.

philosopher, Professor Hans Jonas of the New School of Social Research in New York, emphasizes his humanistic approach to reality by stating (9, p. 119), "It is I who let certain 'messages' count as 'information' and as such make them influence my action."

Both of these statements point to one fundamental question that has been posed by man from time immemorial: Is man free to choose his own destiny and thereby circumvent the forces of nature?

In the context of the current impact of science-technology on society, it appears that man *has* freed himself from the uncertainties of natural forces by harnessing to his advantage every conceivable resource—water, land, fossil fuel and atom. But a closer examination reveals that this extreme "exteriorization" has proved to be man's ultimate predicament. Apparently unshackling himself from nature, man has built with pathetic persistence another bond of servitude—a dependency on automation, computers and information retrieval. As Loren Eiseley states so elegantly (5), "This new judgment is an easy one; it deadens man's concern for himself. It makes the way into the whirlpool easier. In spite of our boasted vigor we wait for the next age to be brought to us by Madison Avenue and General Motors. . . . We wait, and in the meantime it slowly becomes easier to mistake longer cars or brighter lights for progress."

Depending on the frame of reference, the definition of man's freedom may vary in essence and degree. As the dominant biotic factor in nature, man is obviously free to make mistakes but, as the spectres of starvation and environmental pollution indicate, he cannot free himself of the consequences. There is, however, a greater freedom—the freedom of conscious judgment—upon which man can rely for resilience and recovery. We shall turn to that momentarily.

One could make the philosophical argument that everything that *is* is natural. Therefore, man's ambivalence regarding his animality and his capacity for conscious reasoning is also in the natural scheme of things. Man's rationality has made him "free" but only within the confines of nature which has a way of dictating terms. From the vantage point of human interpretation, this relationship indeed assumes a uniqueness. In this connection, an enlightening analogy from Hindu religious teaching comes to mind: Tributaries of the same river—the river of Life—juxtapose different landscapes, inspire different responses among observers, and finally merge together to enter one great ocean wherein they lose their individual identity.

RECONCILIATION

In order to identify man and his attributes with one continuous timeless natural process, one has to recognize a gradient in the products of evolution. The same mechanistic model must be applied to matter and mind, instinct and reason, in an ascending order of organic power and function, culminating in man. Simple metabolism, primitive reflex, instinct, complex behavior, intricate neuro-muscular coordination, feeling, reason, imagination—all these are characteristic elements of this gradient. To reflect the thought of Jonas (9), they constitute a kind of "graded omnipresence" in a single scheme.

Much controversy prevails as to whether one set of physical and chemical laws governs the functioning of all levels of organization in life. The old conflict between the vitalists and mechanists of early centuries has assumed an intriguing dimension in modern times. Today, the natural scientists can be divided into two distinct schools of thought—the *reductionists* and *anti-reductionists*.

The adherents of the "reductionist" school appear to base their judgment on one premise: The biochemical pathway of evolution could not deviate from the thermodynamic property of matter which brought about prebiological synthesis of macromolecules in the primal oceanic surroundings (see p. 5). At the risk of oversimplification, it can be argued, therefore, that if the entire genetic information within the ancestral species of an extant animal could be known and fed into a computer, the current species would be predicted in its entirety.

The Nobel laureate biologist, Francis Crick (1, p. 98), maintains that our present knowledge in molecular biology "makes it highly unlikely that there is anything that cannot be explained by physics and chemistry. . . ." In other words, with sufficient understanding of the interplay within the hierarchy of atoms and molecules in a living system, Crick feels that organisms can be described by purely mechanistic models. Wooldridge (20, p. 204), following powerful scientific persuasions, concludes that "the same body of natural law will . . . suffice to 'explain' the formation of a distant nebula, the operation of a television receiver, the growth of a child and the genius of an Einstein." Although he admits that the above statement may contain an "article of faith," Wooldridge is optimistic that a single body of physical law will soon "account for all aspects of human experience . . . ." Schaffner (14) lends his support to this position. He finds complete sufficiency in chemical systems for explaining biological processes even without having to invoke an evolutionary overview.

Generally speaking, "anti-reductionists" do not believe that life can be reduced to simple physical and chemical terms. These scientists do not deny, however, the roles played by basic laws of physics and chemistry in the functioning of a living cell. We return to Polanyi (12) once again. He argues that a living organism represents a hierarchy of organization, each higher level harnessing the one immediately below it, but never reducible to it. In the sequence of voice—word—sentence—style—creative composition, one element crosses the boundary to become another element. A "sentence," for instance, must depend on the assembly and property of "words;" and yet, one cannot reduce the "sentence" to "words" without destroying its design and identity. In an infinitely more complex hierarchy such as we find in living organisms, Polanyi sees "progressive intensification of higher principles" that can no longer be reduced to the chemistry of macromolecules. "Irreducible higher principles are *additional* to the law of physics and chemistry," he insists.

The well-known physicist, Walter Elsasser (6, p. 504), recognizes "phenomena in the organism that cannot be explained in terms of mechanistic functions." Instead, he feels that two principles—one *deterministic* and the other *creative*—must intermingle for life to be manifest. Similar, but not identical, arguments about the "intrinsicality" of life have been made by scientists such as Siu (17), Sinnott (16), and Wigner (19).

At this point in evolution, man's conscious effort is the only thing that can resolve the question of life's reducibility or nonreducibility. This introduces a new problem in the application of *cybernetics*[2] to the complex system of human behavior. The opera-

---

[2] *Cybernetics* refers to the science dealing with the functioning of automatic machines (from the Greek, *kybernētēs*, steersman or pilot).

tion of highly sophisticated machines, such as computers, depends on the relay of signals and is regulated by feedback mechanisms. Unlike these systems, the human machine does not respond fully to feedback; it can and does exercise deliberate efforts that will defy a purely mechanical model. Due to the historicity (17) of human intelligence and perception, the extension of the cybernetics of inanimate systems into human propensity may be totally premature. The massive beauty of the special science of cybernetics, however, has seduced the unsuspecting human mental activity—the "epiphenomenon of material processes in the brain" (9, p. 110). Is it possible that man's evolution, unlike the evolution of other organisms, has assumed an extraordinary dimension that permits the evolved to *control* its own evolution, governed by a new set of laws heretofore unknown?

FEAST OF KNOWLEDGE AND FAMINE OF WISDOM

For all practical purposes, it is not biological evolution but the functioning of the human mind that constitutes the lowest common denominator of life. We must hasten to note, however, that these two factors are not mutually exclusive; after all, the mind is the epitome of biological evolution, as the forthcoming discussion would attempt to clarify.

Starting with the above premise, knowledge and wisdom mastered by the human mind will singularly determine if the drama of man and nature may end in tragedy or in reconciliation. Together, the terms "knowledge" and "wisdom" give the impression of a contrast, but this is not necessarily the case. It is true that an endowment of factual knowledge may not guarantee wisdom. Conversely, wisdom, "the power of true and right discernment," does not presuppose possession of knowledge. A simple analogy may clarify this distinction further: The fact that the sun rises in the east and sets in the west is a part of the total pool of information on which we all depend. The inspiration that poets, philosophers and naturalists derive from the riot of splendid colors at sunrise or sunset, however, does not depend on our knowledge of dispersion and refraction of light. Here, the knowledge becomes too obvious to be suggestive.

In 1883 Francis Galton ventured beyond the knowledge that the human body was made of billions of cells and wondered about the "conscious life of the man as a whole." Galton wrote, ". . . our part in the universe may possibly in some distant way be analogous to that of the cells in an organized body, and our personalities may be the transient but essential elements of an immortal and cosmic mind."

· Curiously, this historical anecdote gives our topic of discussion greater currency.

At the risk of repeating what is too obvious by now, we may turn once again to the feast of knowledge which has been in full swing (p. 95). John Platt (11, p. 1115) estimates that "in the last century, we have increased our speed of communication by a factor of 10^7; our speed of travel by 10^2; our speed of data handling by 10^6; our energy resources by 10^3; our power of weapons by 10^6. . . ." With a wealth of knowledge we have constructed impressive buildings, augmented industrial output, maximized food production and controlled lethal diseases.

Yet, seen against the backdrop of his extraordinary biological achievement and psycho-social potential, the dilemma of modern man's own future seems extremely incoherent. Herein lies the greatest paradox of all. The lunar landing of man overshadows starvation on earth. False military and economic security deadens our sensitivity to basic human needs. The dream of a single planetary society, finding expression in such magnificent organizations as UNESCO and UNICEF, becomes a nightmare as nations continue to be torn by conflicts from within and without. Finally—and most importantly—the "power of exported death control" (p. 101) deceptively increases the hope for quality living as teeming multitudes deplete Mother Earth to a gnawing bareness.

Impressed with the above paradoxes, the author's student wrote in a term paper: "The vision is not a pretty one. And yet, it is impossible to condemn man for doing what evolution prepared him to do."[3] Just what did evolution prepare man to do? By the strict definition of the term, evolution does not *prepare* anybody to do anything; it merely confers a predisposition. This predisposition, under man-made circumstances, is capable of producing Aristotle,

---

[3] Miss Sally G. Damon, class of 1970, Smith College, Northampton, Massachusetts.

Beethoven, Shakespeare, Tolstoy, Lincoln or Einstein. It is also capable of the "Hiroshima" massacre and attritional genocide. It has created individuals and institutions. When identified with collective knowledge, we call it our property; when it becomes our collective wisdom, we call it our virtue.

RATIONALIZATION OF HINDSIGHTS

A crisis such as a war or an epidemic is so devastating and dramatic that it warrants plans for reconstruction of the community and rehabilitation of the sufferers even before the crisis in question is over. We have witnessed this to be true in the post-war reconstruction of the forties — the crash program that led to rapid improvement of technology, drugs and food. If, on the other hand, a crisis creeps up on us without the fanfare of war or epidemic, we are caught unguarded and desperate. It appears that the ecological catastrophe that mankind faces today *is* such a crisis.

We may attempt to put the profile of this crisis in line with a socio-economic viewpoint. The inanimate power of mass production assumes an inherent anonymity. Corporate decisions seek profit or counter loss by increased efficiency and curtailed cost of production. The failure of such a system takes a relatively long time to become manifest. It proceeds under cover of apparent progress and prosperity until overpopulation and pollution — the warning signs of a communication breakdown — finally reach the doorstep of the decision-making units of our society. In this process, we may fail to recognize the villain, or recognize it only when it is too late. One can say with certainty that this "delayed kill" never occurred in the agrarian style of life where the feedback of success or failure of man's actions arrived with unmistakable promptness; the test of "fitness" was never postponed.

When we speculate on the direction of man's evolution, two provocative views emerge. As a part of nature, everything man does — right or wrong, wise or unwise — is also a product of nature. If man-made pollution chokes man to death and his lack of communion with nature erodes the soil, denudes the forest and wipes out rare animals, then his biological evolution will simply have to make a fresh start. It would mean that his *defeat* in nature necessarily follows a transient *victory*. Since nature is amoral, the morality of man's victory or defeat has no meaning outside human parameters. Therefore, it appears that the morality itself must evolve through the painful trials and tragic errors of man, over a long period of time. Whenever man creates something rationally, he cherishes it until he destroys it irrationally and comes to regret it. The cycle goes on. Admittedly, this view of man's evolution is fatalistic.

The other view portrays man's existence as a flash on a cosmic time-scale. It is, however, a very significant flash, heralding the dominance of convergence over divergence. The predisposition conferred on man by evolution potentially insures his ability to accept and overcome his individual isolation—"the passing whiffs of insignificance" as A. N. Whitehead puts it—and to realize his part in an existence more enduring. His identification with social causes, groups and institutions generates a cohesive purpose that, according to the famous biologist, Dobzhansky (3), has made human history different from biological history. "Man plans and has purposes. Plan, purpose, goal, all absent in evolution to this point, enter with the coming of man and are inherent in the new evolution which is confined to him" (15).

The distinguished paleontologist and Jesuit Father, Teilhard De Chardin (18), in an extraordinary synthesis of science, metaphysics and theology, asserts that the whole universe is one vast evolving system in which man occupies the growing point. Chardin maintains that the "noosphere," the sphere of mind, is superposed on the *biosphere*, the sphere of life, and that this confluence itself is evolving toward a sort of superconsciousness. Huxley (8), who essentially agrees with Chardin, presents a naturalist's view of the same process and envisages a "progressive psycho-social evolution" comparable to the "noosphere."

In view of the foregoing discussion, it appears that in man "evolution is at last becoming conscious of itself." The future of human evolution will lie *not* in unpredictable, random, protracted, biological evolution, but in studied, predictable, conscious human judgment. Earthbound and hurtling through dark, endless space, man is riding a delicate craft. The escape-velocity of evolution that put man in orbit around the non-humans as their dominant overseer can be sustained only as he becomes human. Does this require a "divine spark," a recognition of the Fatherhood of a seemingly unknowable God and the illusive Brotherhood of man, or a greater symbiosis with concomitant sacrifice of personal freedom? We only know that man alone can resolve that question with introspection, compassion and love of life.

AFFIRMATION: A PROGRAM OF ACTION

How do we restore the plundered earth and make peace with nature? We could summarize the answer in one sentence: Man's short-term effort of expediency must promptly change to a long-term strategy for survival. Without a program of action, this prescription obviously amounts to an assembly of empty words and may be mistaken for a cure. Rienow and Rienow (13) concluded their powerful dissertation with concrete suggestions for "prolonging our moment in the sun." A similar attempt has been made by Ehrlich (4) with more intense political overtones. In keeping with the emphasis in this discourse, we shall now elaborate upon several closely related programs of action.

In the ultimate analysis, the crisis we face is world-wide and *not* limited to any particular country. As Ehrlich puts it so graphically, the wishful thinking that the population explosion is a problem only of underdeveloped countries is like warning a fellow passenger, "Your end of the boat is sinking." We live in a pluralistic world of many nations with diversified cultures— affluent and poor, traditional and modern. Millions of people have been denied the luxury of a "plan" for survival since they are already stripped down to their basal instinct of fighting death. What constitutes a minimum standard of living varies considerably from one country to another. Forms of government differ substantially as do their influence on the attitudes and values of the people. Religious adherence affects food habits, reverence for life, and regulation of human reproduction.

Can society or an institution induce man to function better than he would if he were left to his own decisions? The United Nations Organization (UNO), in which man's collective conscience is presumably embodied, can and must tackle that question today. Without attempting to obliterate cultural diversity among nations, this organization must begin to bring the concerns of mankind into sharp focus. A suggestion has been made for an *International Ecological Year* fashioned after the successful *International Geophysi-*

cal Year (1957–58). Under the auspices of the UNO and UNESCO, experts from various disciplines should gather and formulate basic guidelines on (a) the population problem, (b) pollution of the environment, and (c) international cooperation in the conservation of natural resources. Above all, they must put forth vital efforts to rehabilitate the damaged image of man. Some of these activities are in progress already, but with little enduring coordination. We should be anxious to create and maintain a renewed mandate of legitimacy for this congress of nations. Regulations on "International Waters" and "Outer Space" as well as sharing of transcontinental communication satellites are reminiscent of the spirit that must be sought to cope with our current ecological crisis.

Needless to say, the world's citizens must be made aware of this mammoth effort with relentless feedback and educational programs. Children of all nations must grow up with a distinct pride that there are sensitive human beings who form policy and plan together to safeguard their future and the future of this planet.

The world-wide nature of the ecological predicament does not preclude national responsibilities, but reinforces them. Coordinating their efforts with international operations, major industrial nations must provide the necessary leadership. Let us examine the vital roles that the United States can play in this connection.

First of all, the government must unequivocally commit itself to fostering an environmental orientation among its citizenry. Obviously, this can be accomplished only by massive educational redirection and promotion of new attitudes and values.

Nothing less than the proven power of mass media can fully convey all aspects of the eco-catastrophe to individual citizens. The complacency of the viewing public can be constructively eliminated as and when "ecology" becomes a household word, with the stigma of "bandwagon" removed from it. This enterprise may initially require a partial sacrifice of profit on the part of the networks as well as subsidies from a determined government and forward-looking industries.

The exponential increase in scientific and technological activities have come to demand a new brand of education—particularly science education—in our high schools and colleges. The cart has been too long placed before the horse. Absolute expertise in methods and tools has been given preeminence over basic education that provides a quality of resilience in crisis. Before it is too late, an ecology-oriented core program must be initiated in our *entire* education system, supported by federal funds for both instructional research and curriculum development. The new order of education will not be a substitute for traditional liberal education and professional training. On the contrary, it should be designed to introduce fresh elements of urgency and sophistication.

The educational program in question is necessarily interdisciplinary. The problem of understanding man's place in the sun entails scientific as well as socio-economic, political, philosophical and religious questions (10). Realization of the need for this synthesis of knowledge is one of the most significant phenomena in academic life today. As the renowned scientist, Bentley Glass (7, p. 90), presents it: "Here we may seek and join the timely with the timeless, the socially relevant with the eternally true, the goals of man with his status in the universe of which he is indeed so small a part."

The right to breathe clean air or to terminate an unwanted pregnancy bring together questions involving the humanities and science, the technological tools and political decisions. How can citizens "nip in the bud" the unanticipated and untested side-effects of technology — side-effects that, if unchecked, can become a calamity. Obviously, the answer lies in relentless political action. In view of the tragic consequences of any further delay, the most important rallying points for young and old alike should be: (a) population control; (b) elimination of highway construction in favor of rapid transit systems; (c) rigid enforcement of pollution control standards; (d) boycotting all non-returnable containers and fostering "habits of recycling"; and finally, (e) creation of well-informed active citizens' groups that can coordinate with similar organizations, nationally and internationally.

One last point must be made with respect to our "program of action." The job of restoring the plundered earth must not be allowed to become a crusade grown out of righteous indignation. It is a salvation that man has to find through his collective conscience. No individual, no politician and no industry should be written off in this endeavor. It must be a joint enterprise of compassionate persuasion and reconciliation — a new brand of evolutionary humanism.

REFERENCES

1. Crick, Francis, *Of Molecules and Men*. Seattle: University of Washington Press, 1967.

2. Davis, Harold L. "Objectivity in Science—a Dangerous Illusion?" *Scientific Research 4* (McGraw-Hill's News Magazine of Science) (April 28, 1969).

3. Dobzhansky, T., "An Essay on Religion, Death and Evolutionary Adaptations," *Zygon 1* (1966).

4. Ehrlich, Paul R., *The Population Bomb*. New York: Ballantine Books, Inc., 1968.

5. Eiseley, Loren, *The Firmament of Time*. New York: Atheneum Publishers, 1960.

6. Elsasser, W. M., "Acausal Phenomena in Physics and Biology: A Case for Reconstruction," *Amer. Scientist 57* (1969).

7. Glass, B., *The Timely and the Timeless*. New York: Basic Books, Inc., Publishers, 1970.

8. Huxley, Julian, *Evolution in Action*. New York: Signet Science Library P2560, 1957.

9. Jonas, Hans, *The Phenomenon of Life*. New York: Harper & Row, Publishers, 1966.

10. McClary, A., "Science-Technology and Society," *Bioscience 20* (1970).

11. Platt, John, "What We Must Do," *Science 166* (November 28, 1969).

12. Polanyi, Michael, "Life's Irreducible Structure," *Science 160* (June 21, 1968).

13. Rienow, Robert and Leona T. Rienow, *Moment in the Sun*. New York: Ballantine Books, Inc., 1967.

14. Schaffner, K. F., "Chemical Systems and Chemical Evolution: The Philosophy of Molecular Biology," *Amer. Scientist 57* (1969).

15. Simpson, George G., *The Meaning of Evolution: A Study of the History of Life and of Its Significance for Man*. New Haven, Conn.: Yale University Press, 1949.

16. Sinnott, Edmund W., *Matter, Mind and Man: The Biology of Human Nature*. New York: Harper & Row, Publishers, 1957.

17. Siu, R. G. H., "The Phoenix," *Bioscience 14* (1964).

18. Teilhard De Chardin, Pierre, *The Phenomenon of Man*. New York: Harper & Row, Publishers, 1959.

19. Wigner, Eugene P., *Symmetries and Reflections*. Bloomington: Indiana University Press, 1967.

20. Wooldridge, Dean E., *The Machinery of Life*. New York: McGraw-Hill Book Company, 1966.

SUPPLEMENTARY READINGS

Bronowski, Jacob, *The Identity of Man*. Garden City, N.Y.: Natural History Press, 1965.

Dobzhansky, Theodosius, "Changing Man," *Science 155* (July 7, 1967).

Kuhns, W., *Environmental Man*. New York: Harper & Row, Publishers, 1969.

Mayer, J., "Toward a Non-Malthusian Population Biology," *Columbia Forum* (Summer, 1969), Columbia University.

Morison, R., "Science and Social Attitudes," *Science 165* (1969).

Simpson, George G., *Biology and Man*. New York: Harcourt, Brace and World, Inc., 1969.

White, Lynn Townsend, Jr., "The Historical Roots of Our Ecologic Crisis," *Science 155* (March 10, 1967).

INDEX

Abiotic components:
 defined, 15
 interaction with biotic, 42
Adaptation:
 fundamentals, 44-46
 to light, 61-66
 to moisture, 46, 48-55
 to temperature, 55-61
Affluence, 102, 105
Agriculture, 96, 98
Animals:
 and ecosystem, 66
 homeothermic, 58
 kingdom, table, 40
 and light, 64
 poikilothermic, 58
 photoperiod, 65
 phototaxis, 64
 and temperature, table, 60
 and water, 50, 52
Animals, adaptations of:
 to light 64-65
 to moisture, 50, 52
 to temperature, 58-61
Anti-reductionists, 124
Ape-man, 91, 92
Arnon, D. I., 7
Atmosphere, 8, 19-24
Atomic age, 98
Automobile exhaust 110

Bacteria, 32, 33
Bates, Marston, 91
Behavior:
 animal, 72-82
 patterns, rhythmic, 69, 70
Biological clock, 70-71
Biological magnification, 112
Biosphere:
 definition, 24
 and sphere of mind, 128
 man predominant 115
Brain, 62, 91, 92
Brown, F. A., quote, 72
Browning, Robert, quote, 119
By-products, recycling, 114

Carbon dioxide, 5, 7
 cycle, 19-22
 in atmosphere, 110
Carson, Rachel, quote, 1, 111
Chemical energy, 7, 24
Chlorophyll, 7, 8, 28

Communication, and evolution, 94, 95
Conservation, 105, 107, 108
 national responsibilities, 130
 responsibility diffused, 106
Crick, Francis, quote, 123
Cropland, depletion of, 106
Cybernetics, 124, 125
Dart, Raymond, 92
Darwin, Charles, 88
Death rates, 98, 101, 102
DDT, 112
Diapause, 61
Dobzhansky, T., quote, 128
Earth:
 atmosphere, 4
 "carrying capacity," 98
Ecological catastrophe:
 crisis, coping with, 130
 feedback prevents "delayed kill," 127
Ecology:
 core program, 131
 crisis world-wide, 129
 man's dominance of, 95
 new attitude vital, 130-31
 and urban blight, 114
Ecosystem, 66, 67, 105
Ehrlich, Paul, quote, 114, 129
Eiseley, Loren, quote, 122
Elsasser, Walter, quote, 124
Entropy, 6, 9
Environment:
 adaptations to, 46-65
 exploitation of, 106
 succession in, 66-69
Environmental quality, 109-15
Environmental resistance, 96-98
Evolution, 4-9, 123-28
 and adaptation, 44, 45
 biologic vs. cultural, 95
 man, 94
Feedback:
 and equilibrium, 10
 and human machine, 125
Fertilizer, nitrate, 111
Fertilizer, organic, 114
Food:
 and animals, 42, 76-78
 energy, 8
 production, 96, 103-4
 supply and man, 96, 98, 101
 world needs, 103-4

134

INDEX

Food chain, basic law, 104
Food-web, and DDT, 112
Forest land, 106
Forests, 50–52
Fossil fuels, 8, 22, 98

Galton, Francis, quote, 126
Garbage increase, 112
Genetic endowment, 45
Glass, Bentley, quote, 131
Gray, William D., 103, 104
"Greenhouse effect," 110
Gross National Product, 102

Habitat and adaptation, 46
Heat, 28–30
Highway construction, 106, 131
Highways, land required, 107
Homeotherms, 58
Homeostasis, 9
Human evolution, future, 128
Human society, urban blight, 114–15
Humans, relative number, 101
Hunger, world, 103
Huxley, Julian, quote, 128
Hydrosphere, 17–19

Industrial economy:
    abuse of environment, 109
    change to agrarian, 103
    irony of, 98
Industrial Revolution, 98, 102
Industry:
    and air pollution, 110
    current boom, 105
    profit-motivated, 106
    and water pollution, 111
Insecticides, 98, 102
Instinct, 72, 78, 81
International Ecological Year, 129
International Rice Research Insitute, 103
"International waters," 130

Jackson, Daniel F., 112
Jonas, Hans, 122

Krakatoa, 69

Lake Erie pollution, 111
Leopold, Aldo, quote, 106
Lewis, G. N., quote, 6
Liebig, Justus von, 5
Life:
    anti-reductionists, 124
    deterministic view, 124
    mechanistic view, 4, 124
    reductionists, 123
    vitalistic view, 4, 39, 46
Light, 61–66
Lithosphere, 15–17
Living system, 3–5, 8, 123

Malnutrition, problem of, 103
Malthusian prophecy, 102

Mammals, definition, 87–88
Man:
    alters environment, 95
    ambivalence, 122
    in crisis, 127
    control own evolution?, 125
    destiny, 122
    dilemma of modern, 126
    dominant overseer, 128
    and evolution, 94, 123, 127
    exploitation by, 95
    freedom of judgment, 122
    knowledge and wisdom, 125
    as natural resource, 115
    and nature, 95, 121–29
    paradoxes of modern, 126
    rationality, 122
    reliance on machines, 122
    strategy for survival, 129
Marquis, Don, quote, 107
Mass media and eco-catastrophe, 130
Mass production, 96, 102
    inanimate power, 127
    inherent anonymity, 127
Marine environment, 34
Mead, Margaret, quote, 105
Mechanistic models, 123
Muir, John, quote, 13
Mumford, Lewis, quote, 115

Natural resources, 96
    conservation, 108
    depletion of, 105–9
    exploitation, 98
Natural selection, 38, 44, 45
    and adaptation, 74
Nature:
    amoral, 128
    dichotomy of man and, 87–88
    and man, 115
    peace with, 129
New School of Social Research, 122
Nitrogen cycle, 20
Noosphere, 128
Nutrition, adequate, 103

Ocean, food source, 104
Organisms:
    adaptations of, 50, 52, 58–61, 64–65, 69–72
    analytic profile, 30–42
    endogenous rhythm, 70, 71
    environment, 46, 70–71, 105
    hierarchy, 124
    and mechanistic models, 123
    navigatory accuracy, 71
    and temperature, 56
Overpopulation, 114
Oxygen, 5, 8, 22–24
    atmosphere, 8
    biochemical need, 111

Oxygen (cont.)
  in ecosystem, 66
  supply threatened, 112
Pesticides, 112
Phosphate pollution, 111
Photoperiod, 65, 69
Photosynthesis, 7
  carbon dioxide cycle, 22
  chlorophyll, 27
  energy, 56
  food-web, 61
Phototaxis, 64
Phototropism, 62
Plant kingdom, *table*, 35
Plants, 38-40
  desert, 50
  forest, 52
  and light, 61-64
  and temperature, 56-58
  and water, 48, 50, 52, 55
Platt, John, quote, 126
Pleistocene epoch, 91, 92
Poikilotherms, 58, 61
Polanyi, Michael, 121
Pollution:
  air, 109, 110
  enforced control, 131
  environmental, 105, 130
  solid waste, 112
  water, 111
Population:
  adaptive features, 45
  control, 101, 131
  expanding, result, 105
  explosion, 98, 108, 129
  interaction, *table*, 47
  level, 11
  problem, 130
  uneven growth, 102-3
Poverty, 102
Preadaptation, 45
Priestly, Joseph, quote, 7
Productivity, world, 103
Program of action, 129-31
Progress, 126, 127
Progression in ecosystem, 67
Protein to enrich diets, 104
Psycho-social revolution, 128
Radiant energy, 25-30
Rapid transit, 131

Raw materials, limited, 103
Reductionists, 123, 124
Resources, non-renewable, 102
Rhythm:
  annual, 71
  circadian, 71
  endogenous, 69-72
  exogenous, 71
Rockefeller Foundation, 103
Rowan, William, 65
Rubbish, composting, 114

Sauer, Carl O., quote, 96
Schaffner, K. F., 124
Science education, 130
Simons, E. L., 88
Solar energy, 7, 25
Solar radiation, 6, 25-30, 56
Starvation, 122
  and lunar landing, 126
Succession, 66-69
Survival:
  basic assurances for, 74
  in ecosystem, 66

Technology, 105
  side effects, 131
Teilhard De Chardin, Pierre, 128
Teleology, definition, 44
Temperature, 55-61
  in adaptation, 46
  survival limits, *table*, 60
Territoriality, 76
Thermodynamics, laws of, 6
Thermoregulation, 55-61
Thurman, Howard, quote, 85
Tidal rhythms, 71
Tools, making, 91, 94, 95

United Nations, 112, 126, 129-30
Urban blight, 105, 114-15

Vitalists, 123

Water, 17, 19
  in ecosystem, 16
  pollution, 110, 111
  role of abundance, 50
  usable and use, 107
Whitehead, A. N., quote, 128
Wohler, Friedrich, 5
Wooldridge, Dean E., quote, 123, 124
World Health Organization, 102